BRUCE DIXON & SUSAN EINHORN

#Anytime Anywhere Learners

A BLUEPRINT FOR TRANSFORMING WHERE, WHEN, AND HOW YOUNG PEOPLE LEARN

Authors: Bruce Dixon and Susan Einhorn
Publisher: Anytime Anywhere Learning Foundation
Cover Design: Stojan Mihajlov
Cover Images: © Shutterstock and Fotolia
Layout Design: Alejandro Natan and Atanaska Mirkova
Website Organization: Justina Spencer
ISBN: 978-0-692-58369-2 (pb)

Anytime Anywhere Learning Foundation
The goal of the Anytime Anywhere Learning Foundation (AALF) is to ensure that all children have access to unlimited opportunities to learn anytime and anywhere and that they have the tools that make this possible. To achieve this, AALF provides a range of support to help schools develop visionary leadership and knowledgeable, innovative educators. AALF is a 501(c)3 non-profit organization.

100% of the profits from the sale of #AnytimeAnywhereLearners will be used to support the work of the Anytime Anywhere Learning Foundation.

© Anytime Anywhere Learning Foundation, 2016. All rights reserved.
Published by the Anytime Anywhere Learning Foundation

Unless otherwise indicated, all images are used under license from Shutterstock.com.

Every effort has been made to accurately reference every quote and text included in this book; however, we would ask that you advise us of any errors in this regard, and accept our apologies for any oversight.

Table of Contents

Preface 5
Why You Need This Book 7

Part I - Three Essential Elements 19

Essential Element 1 - The North Star: Vision, Mission, Goals 21
Understanding the Context 26
(Re)Define Your Vision 38
Clarify Goals and Policy Priorities 53
Evaluate - Continuously! 60
Essential Element 2 - What's Now Possible: Teaching & Learning
for Contemporary Learners 65
Explore Contemporary Learning 67
Embrace New Roles for 21st Century Educators 84
Design a Modern Learning and Teaching Environment 96
Essential Element 3 - Cultures of Change 103
Build a Change Culture 105
Implement Professional Learning Strategies 114
Develop Funding Strategies for Equity & Sustainability 125
Build Community Support 131

Part II - Implementation 135

The Implementation Phase 137
Conduct a Readiness Assessment 141
Consider Implementation Options and Project Plan 144
Choose Your Devices, Core Tools, and Apps 147
Plan Your Infrastructure 151
Prepare the Budget 154
Establish Critical Partnerships 156
Manage Support Services 159
Create Effective Policies 164
Liaise with Parents and Community 167
Deploy 170

Part III - Tools and Resources 173

1:1 Checklist 175
Articulate Your 1:1 Vision 184
Designing 1:1 Pedagogy Statements 185
AALF Policy Decisions 190
Infrastructure Questions 191
Frequently Asked Questions About 1:1 195

Preface

The story behind this book is a very long one, but here it is in very short form.

Tens of millions of young people around the world have their own personal computing device today because of the extraordinary vision of one man, Seymour Papert, who, together with his colleague Alan Kay, believed that one day each and every young person would have access to *"an instrument whose music is ideas"*.

Image credit: © Alan Kay

From that genius vision they had in 1968, it was to take more than twenty years before it became a reality for the 9 and 10 year-old girls at Methodist Ladies' College in Kew, Victoria, Australia. The '90's saw their powerful idea seeded across schools in Australia and then, as it launched in the US in 1996, lead to the creation of the **Anytime Anywhere Learning Foundation** (AALF).

The Foundation was established to provide thought leadership and advocacy for what we believed this vision could make possible for young people, at a time when the vast majority of people doubted both the value and the viability of the initiatives being undertaken. It was also created to support the energy and spirit of those pioneering schools and to provide a collective knowledge space where people could share their experiences and ideas.

Much of this was achieved through the evangelism that came from a series of eight Anytime Anywhere Learning Summits that started in the late '90's, run in partnership with Microsoft and other industry participants; several years later an event partnership with Apple in Australia ultimately led the government there to provide a million students with their own laptop.

As more people came to understand the expanded learning opportunities that were possible through these '1:1' initiatives, more schools embraced Papert's vision; but early on, it was obvious that there was a critical need for a roadmap to guide schools through the steps required to ensure effective implementation. At the dawn of the new millennium, this resulted in a partnership, initially with the Queensland (AUS) Ministry of Education, that gave rise to what became the *21 Steps for 21st Century Learning*® workshops.

Since then, the workshops have expanded and been attended by thousands of school, district, and policy leaders across more than

Introduction

45 countries. All the while, the content has been continually revised to ensure it stays current and relevant and provides the most beneficial support to teachers, school leaders and, most importantly, students. In recent times, another partnership with Microsoft has enabled AALF to run a series of masterclasses for workshop leaders across an additional 20+ countries.

Which brings us to this book, which we believe encapsulates the best of what we have learned, together with links to the best videos, websites, readings, white papers, and references that we feel will support your school as you embark on this journey.

This is a roadmap for that journey; it is the result of thousands of hours of school visits, workshops, conferences, and long conversations with the extraordinarily courageous and committed educators and educational leaders who believe in Seymour's vision of so long ago.

We would especially like to thank Adam Smith, who not only helped in polishing this book, but who, as one of the first 1:1 teachers in the world, had the courage and pedagogical wisdom to bring Seymour's vision to the students at Methodist Ladies College Australia.

This book is dedicated to every one of them and to the father of educational computing, Seymour Papert.

Susan Einhorn and Bruce Dixon

Why You Need This Book

This book is a guide for leaders of schools and districts that are seeking to create a modern learning environment built on the simple proposition that young people will have unprecedented opportunities for learning when they have unlimited access to their own personal portable computer. Commonly many refer to this as 1:1 and while this can at times infer a focus on simply providing a device to a child, the real agenda is about the learning that it now makes possible.

It focuses on building learning environments that unleash the potential and thinking, doing, and creating opportunities that happen when technology is ubiquitous, personal, and wisely used. Its goal is to help you plan, design, and implement your initiative with this goal in mind.

The framework outlined in this book is based on the experiences of thousands of 1:1 schools around the world over the last 25+ years. In the interim, not only individual schools and districts have moved to 1:1, but whole states and, in some cases, countries, have invested in providing their students access to personal, portable digital devices. Information, research, and feedback from these initiatives have helped clarify the essential components for getting the best outcomes for learners today.

Rather than viewing the framework as a step-by-step prescriptive methodology that has been 'laboratory' developed and tested, it should be understood as a distillation of the experiences and knowledge accumulated over these years - essentially an examination of initiatives around the world that has been revised over time as more and more schools and districts have become 1:1. (In a similar vein, the white paper **A Policy Agenda for a 21st-Century Education**[A1] reviews and analyzes 1:1 policy development worldwide over the last 25 years.)

This is not a textbook or a checklist. Nor is it an Ikea®-like set of instructions to build an easy-to-assemble 21st century school. This guide delves deep with the expectation that you and your team understand a transformation of this size takes effort and an investment of not only money, but time and commitment to an outcome that will be reflected in the learning experiences of your students, *not* simply as a number that reflects device ratio or density.

This is not a book to read from start to finish; it is meant to be a **working** document, something that you can build on, add to, refer to and use as a discussion starter or provocation for critical conversations at various times with your school community.

It contains information, essays, resources, and a collection of questions and recommended actions to help you, as school or district leader, determine what you need to know and do at each step of the 1:1 planning and development process in order for you to be able to **make critical decisions, know the strategic questions to ask, and have realistic expectations about what the outcomes will be and what will be required of both district personnel and school leadership.**

Consider each resource, question, and action carefully since the theoretical and practical knowledge you develop will provide the foundation you need to make your vision a reality.

> **Important Information on Links**
>
> The following protocol has been put in place to ensure all links are kept up-to-date. A letter-number label is included in superscript after each resource, which is the link identifier. To access the resource online, go to any browser and type: **www.AnytimeAnywhereLearners.com/?** substituting the link identifier after the resource title for the question mark. For example, if the identifier after the resource is A5, use the URL www.AnytimeAnywhereLearners.com/A5
>
> (Note: link identifiers are not case sensitive.)

What to Expect

The **#AnytimeAnywhereLearners** framework is divided into three parts.

Part I focuses on three **Essential Elements**, each of which helps build the foundation required for subsequent elements. These Essential Elements do not all start at the same time. Rather;

- they are undertaken in a strategically organized order,
- the launch of each subsequent element starts only when the preceding element is well under way,
- the work for an element continues even as the next element comes into play.

In this part, you'll explore **three questions**:

1. **What is the compelling case for change?**
2. **What does ubiquitous access to technology now make possible for teaching and learning?**
3. **How might you best lead this inevitable shift?**

Only once you've answered these questions, are you then ready to determine how best to manage implementation. Designing infrastructure, purchasing, setting up support services, and other key actions are subsequently covered in **Part II**.

Part III then contains tools and resources you will use throughout the planning and implementation of your initiative.

Notice that this framework starts by asking you to carefully determine what the reasons are for beginning what will be an ongoing, often challenging process that has the potential of creating and enormously expanding learning opportunities for all your students. As you do this, you explore what these opportunities may be and the impact they may have on all your current expectations about the role of school. This process requires a clear understanding of what it takes to build a culture in which change for the better is encouraged, supported, and accepted. Only once all these ideas and strategies are explored does it becomes possible to determine the technical elements that will best support your vision and goals.

There are times you may want to jump ahead, to begin purchasing devices, software, or infrastructure. For many this would seem the place to begin, and they may try to convince you to skip the earlier elements. Resist!

Over two decades of experiences have shown that the place to start is at the beginning – with a clear understanding of why you are undertaking this shift, how you envision learning will be in your school in five to seven years, and what this will mean in terms of learning and teaching.

If those ideas are not in place before you purchase devices, you may find yourself shaping your vision to your device capabilities instead of the other way around. The problem with 'device-first' projects is that they run a huge risk of being merely device deployment initiatives, not learning initiatives.

What You'll Find

This guide describes a framework for knowledgeable action. It starts with the big picture, the helicopter view, and moves down to the nuts and bolts.

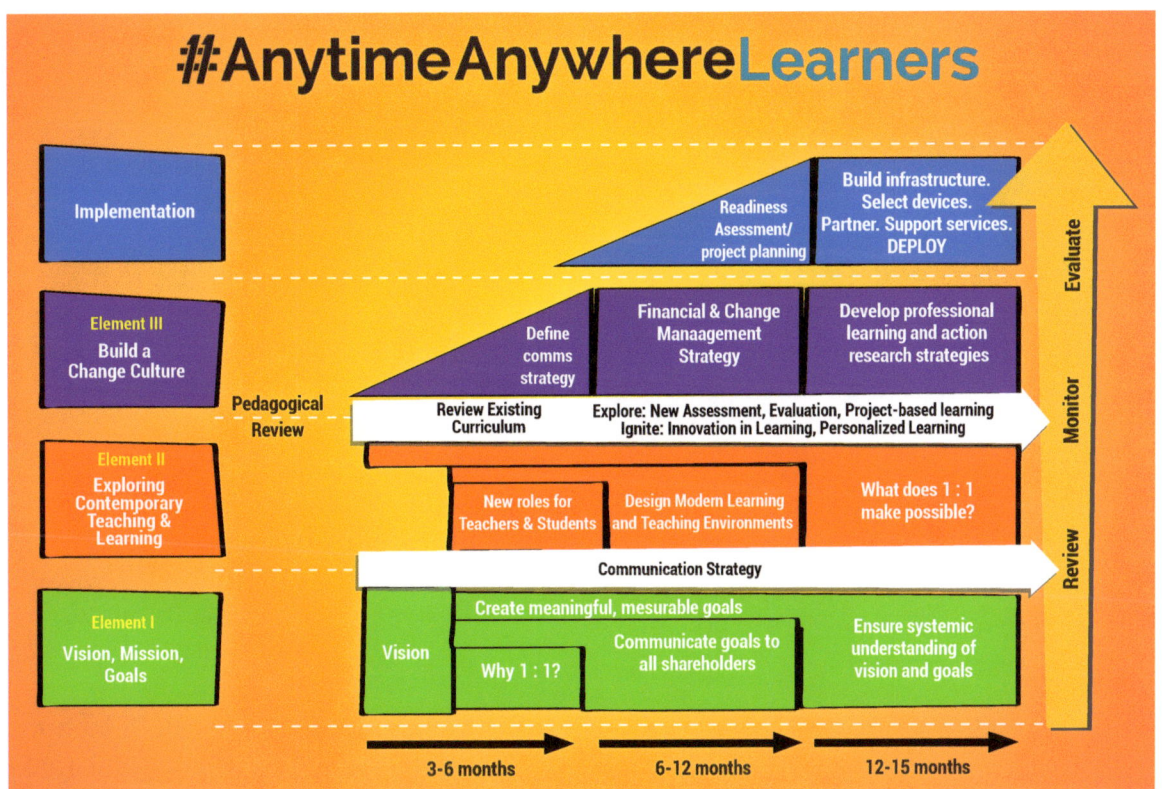

Introduction

The **Three Essential Elements** and **Implementation** parts of this book are divided into several chapters and focus areas. As you work through this book, you will need to complete each section based on *your* vision, *your* goals, and *your* requirements.

- Each chapter consists of two sections. The first introduces the key principle, the main ideas around it, and the role it plays in designing and implementing a successful initiative.

- The second section contains **Questions & Actions**. Note, this section calls for *Actions*, and does not provide *Answers*, since the answers for these critical questions are not in this guide, but **must be determined by you and your team**.

The research, thinking, and discussion that is required to answer these questions are important steps in developing your initiative plan. It is important to neither ignore these nor skip over them. You may choose to assign different questions to different members of your team but make sure to discuss them with the group as a whole.

Within this section, you will also find links to recommended videos, articles, research, and tools. In particular, a number of chapters include links to videos created specifically for this framework. They will provide valuable guidance and should be considered essential resource materials.

Take the time to review the other resources as they will provide opportunities to explore a range of ideas, perspectives, and approaches that will help you as you determine what you want your 1:1 initiative to achieve.

As you work through the questions, it is important to assemble your research, analyses, and intended actions into a document containing your final overall initiative plan. Plan outline questions are included at the end of each chapter and, as well, a complete plan outline is included on the AnytimeAnywhereLearners website.[A14] If you find you have questions that are not in this guide, but are of importance to your school or district, add them so they become part of your final planning document. As well, add links to any additional resources you use. Your planning document should be a complete and accurate record of the research and actions you undertake throughout the process.

There are many questions to answer and answering them will take time, but, once done, you will be able to begin your initiative knowing you've prepared a very solid foundation.

Finally, this guide also includes a number of **Critical Conversations** - essays, short articles, and other insights and opinions designed to inform and provoke new ideas, new conversations, even disagreement and anger. They provide additional depth and context for many of the topics you explore in this book, as well as important questions that should be considered as you engage in each stage of your initiative. In some cases, you might assign them as background reading to all or some of your team; in other cases, you may see the content as optional or extension reading. Read all of these, some of these, but **not none** of these. They are here to provide a breadth of material to help all of your team members be fully engaged, well informed, and deeply committed.

Final Note

Remember, this is **not** about technology. It is about what the technology makes possible.

Critical Conversation: What is the Role of School in Contemporary Society?

To understand the impact technology can and should have on our schools, it is important to first go back and take a deeper look at exactly what the role of school has been, how that has changed over time, and, finally, begin to consider what the role of school should be today.

Society accepts the existence of school as a necessary, natural part of life. When children turn 4, 5 or 6 years-old, depending on the country, parents, with few questions, surrender them to a team of adults, most of whom they do not know personally and with whom they willingly begin to share the responsibility of their children's upbringing, expecting positive outcomes at the end of 12 or 13 years. School has become so much a part of our life that we rarely explore what we expect or why we assume school will provide it.

We tell our children to participate in this institution because it is what is expected, what is required. It is part of our universally agreed upon culture and a rite of passage. Young children eagerly await that first day of school with high expectations. But what do they expect? What do their parents expect? And what does the community and society that supports school expect? What role do they believe school plays in their children's lives? How does it connect with the role school plays in society?

No discussion of the role of school is simple. Politics, ambition, class systems, religion, as well as benevolent concern for the well-being of the poor, all impacted decisions about what role formal education — School - would play. It would be misleading to draw a straight line from one initial rationale to today's schools to explain its role in a country's existence and our individual lives. Yet some ideas seem to persist, not all of which are compatible with each other, making it difficult to clearly point to a single, clearly defined role. This not only causes a constant re-thinking of curriculum and methods but has resulted in a definition of school success that may not in any way be consistent with our own personal or lifelong understanding of success. Perhaps, this is why so many young people leave school without a clear sense of how to use much of what they have learned or even what to do with their lives.

Today, the education of young people, regardless of their family income or ancestry, is almost universally accepted as the responsibility of government. In the long and often complex history of schooling, a number of different reasons were put forth to justify these government expenditures. The role school played derived from the expected benefits, falling into two general categories: those that focus on the benefits to society as a whole and those that focus on the benefits to each child as an individual. Understanding the role school has and can now play has to be the starting point for any conversation.

Whereas education and learning were part of the life of the wealthy for centuries, designed to provide the necessary skills and knowledge for maintaining the social position that family wealth bestowed and society accepted, School as an institution, especially an inclusive institution provided for all young people, has historically served a number of different societal roles. Among these are social cohesion, developing the habits and knowledge to be informed citizens, the absorption and assimilation of large scale international and internal migration, and economic growth.

Introduction

A second category arose that placed the individual child's needs and goals at the center of School's work. It focused on the development of the individual and specifically formed the core of the progressive learning movements that became popular at various periods, for example, the early 20th century, and the open classroom movement of the mid-twentieth century. According to educator John Dewey, their specific focus was that School's role was to help each child reach her full potential, discover her passions, focusing on the child, not society's needs. Child-centered learning was at its core.

If one point is clear, it's that the line separating these two categories of roles is blurry at best. It would be illogical to assume that if society as a whole benefits, the individual does not. Preparing students for the jobs or industries that will provide for future economic growth of a country means preparing them for potential career opportunities. And a young person who is allowed to reach his full potential would then be best able to participate in and contribute to society. But Schools vision, goals, organization, curriculum, and environment will develop very differently depending on which role is seen as the primary role. And the role is very dependent on the culture of the time and the tools and media available for learning (media matters).

Assimilation, Social Order, and Citizenry, Democracy and the National Culture

Historically, schools served multiple societal purposes. In the US, for example, assimilating immigrants, helping to remove any traces of their distinctive foreign behaviors and beliefs, has been a key driver in building public schools and the school curriculum. As newcomers came en masse to the US, whether the Irish in the early 19th century, the Italians and Eastern Europeans in the early 20th century, or Asians in the mid-20th century, the 'melting pot' of school would turn them all into true Americans. Their goals, perspective, behavior, language, and beliefs would be indistinguishable from those of citizens already populating the country. The corollary to this belief was the hope that schools would be the tool to prevent immigrants from 'destroying' the culture already in place:

> *The consequence was that Europe was pouring in upon our country an increasing tide of her ignorant, superstitious, degraded and oppressed population. Many thought a momentous crisis was at hand, and that something should be speedily done to countervail the baleful influences which appeared to be sapping the very foundations of our institutions. Public attention was naturally turned to our common schools as the palladium of our liberties.* [A2]

Schools were seen as the means of passing along cultural values from one generation to the next. This became especially important with each new wave of immigrants. How were they to understand their new homeland and be able to participate in its life? There were few mechanisms available to help new immigrants assimilate into their new culture, which meant not only learning the language but the cultural traditions each nation holds as most important. Most support organizations focused on charitable work, providing food, clothing, and shelter, but with few resources available to do more. School became the main mechanism to pass along these foundational cultural beliefs.

In New York City, school kept hordes of poor immigrant children off the streets as their parents worked. Not only did School prepare these children for future factory work in support of the goals of the industrial age, it was a means of preserving the democratic principles from which the US was born.

> *Since the Revolution, leading statesmen had expressed their belief that illiteracy was a threat to the high ideals of the new nation, that a democracy with broad suffrage must have an enlightened electorate.* [A3]

In **The Children of the Poor,** Jacob Riis, a 19th century social reformer, wrote, "The immediate duty which the community has to perform for its own protection is to school the children first of all into good Americans, and next into useful citizens." [A4]

This role of school in the US was not only recognized by American educators, but even noted by other countries...

> *American schools have had, as an avowed purpose, the Americanization of children from diverse cultures, races and climates.*
>
> The Plowden Report, England [A5]

As well, schools were seen as the means to keep society orderly. According to Benjamin Rush, a US founding father,

> *I am satisfied that the most useful citizens have been formed from those youth who have never known or felt their own wills till they were one and twenty years of age, and I have often thought that society owes a great deal of its order and happiness to the deficiencies of parental government, being supplied by those habits of obedience and subordination which are contracted at schools.*[A6]

Many educational experts today draw a direct line from the needs of the industrial era to the design and curriculum of schools for the last century and a half. **The perceived need was for a compliant workforce prepared to work in a structured environment and repetitive work performed individually. Even the functioning of school was modelled to be as efficient as a well-run factory, processing students in batches, assessing their progress via quality assurance type of testing.** Many schools used the Lancasterian method of school in which each grade could be divided into smaller (50 student) groups led by monitors (not teachers) who each followed a rigid curriculum in which students were drilled on a set of facts and required to respond exactly as practiced. Any deviation was deemed an error, a failure. All monitors followed the exact same script. School held the information and knowledge and delivered it to their compliant charges. School provided what students couldn't get elsewhere – content, in a very precise form.

The efficiency with which this system functioned resulted in reducing the cost to educate large groups of mainly poor children. It also produced the results required – a workforce for the manual, repetitive labor required by industry. There was no need to question its value to society or attempt to change the system. There were other issues more pressing. School's role seemed to be set.

In England, schools existed in various forms since about 600 AD, most designed to prepare young people for the clergy.

> *However, the only subject taught systematically in the grammar schools was Latin grammar and literature, because the aim of the schools was strictly vocational: to prepare pupils for entry to the Church.*[A7]

By the 19th century, the debate in England about the role of school pitted those who believed it was for industrial training, against those who felt it supported democratic principles and others who believed in the 'moral rescue' of those of the lower class by providing all children with a minimal level of education. What emerged was a curriculum that was "a compromise between all three groups, but with the industrial trainers predominant."[A8]

In a nod to the lives of the individuals involved, education was seen as a means to reduce poverty and school assumed the role of social services provider. School and education were seen as the means to end what later became known as the "cycle of poverty." In addition, the recognition of new roles, such as that of school psychologist, emerged from the social reforms directed at the needs of children that were enacted in the mid to late 19th century in both the

Introduction

UK and the US. Compulsory schooling laws came into effect beginning in the same era, another intervention designed to ensure no young person would be denied an education, with education now recognized as being of equal importance as adequate food, clothing, and medical care (and enforced under the doctrine of *parens patrie*). The institution of school became an instrument in helping to ensure the rights of children would not just be defined, but be open to some form of oversight.

Even today, the UN Millennium Development Goal #2 focuses on achieving universal primary education by 2015, for all children - boys and girls. It recognizes education, a key means to reduce global poverty and increase the healthiness of the community, as a major benefit to society. (Unfortunately, this goal has not been reached.)

Some school leaders like William Henry Maxwell, Superintendent of New York City schools in 1900, bridged the two categories of school roles. A progressive, he strongly believed the focus of schools must be on the individual child, and, according to Diane Ravitch in *The Great School Wars*, he envisioned school as the instrument to take children "from the gutter to the university."[A9] As Ravitch continues, "Though his interest was more in reshaping children than in reshaping society, Maxwell nonetheless placed the public schools on the side of social concern and social betterment."

Shaping the Child

As early as the mid-1700s, attempts were made to shift the role of school from primarily a societal focus, to first and foremost a focus on the child and how she learns. These ideas were influenced by a variety of people, from Rousseau to Dewey and Piaget to more recent educators and thought leaders such as Meiers, Papert, and Robinson, all who focused on the child as an individual, with strengths, weaknesses, passions, and a creative, curious nature. Child-centered learning is often criticized as a less rigorous approach to education, one that assumes not all children are capable of following an intensive prescribed course of study across the range of disciplines deemed to be the cornerstone of intellectual development — the very essence of the educated adult. Its defenders, on the other hand, note that learners, when allowed to pursue personal passions, can be guided to learn more deeply, developing habits of mind that only come about from a more profound exploration of a topic or discipline. This doesn't necessarily mean the work of learning is easy, but it may ignite a strong desire to learn and be a learner. Since it is estimated that today's graduates will have 10 or more different jobs in their lifetimes, being curious about things that matter in one's life, being open to learning more about them, and having a passion for learning will be essential skills that will benefit society as a whole.

Even as they prepared the children of the poor to be useful citizens and workers in the new industrial economic base of the city, the Free School Society's schools in New York --the precursor of the city's public school system--was also determined to help the poor children of immigrants receive some level of education, reflecting a concern for the welfare of the child even as their efforts benefitted society. This was a focus on the child, not as an individual, but for individual needs. By no means, though, was this a child-centered view of the education schools could provide, since these schools were run with Lancasterian precision. That role was to develop only later, particularly once educational philosophers such as John Dewey (in a somewhat anti-industrial philosophy) focused on the child as a passionate, curious learner driven to know more when actively involved in authentic projects.

> The Plowden Report, in describing the aim they proposed for primary school states:
>
> *The school sets out deliberately to devise the right environment for children, to allow them to be themselves and to develop in the way and at the pace appropriate to them. It tries to equalize opportunities and to compensate for handicaps. It lays special stress on individual discovery, on first-hand experience and on opportunities for creative work. It insists that knowledge does not fall into neatly separate compartments and that work and play are not opposite but complementary.*
>
> *At the heart of the educational process lies the child.*[A10]

John Dewey emphasized authentic opportunities to learn by participating in work that benefitted the community as well as the individual child. His emphasis on cooperation, sharing of ideas and support, participating in projects that were of significance to the community and not only an individual child, non-competitive, non-isolated learning is reflected in ideas expressed by educational thinkers today, such as Seymour Papert and even those in the maker movement community. Dewey spoke of developing critical thinking skills as well as creativity, and of helping students make connections between the work they did and social and scientific values. His ideas came at a time of both cultural and political shifts and influenced a number of policies. But, although child-centered learning as a theory and policy was adopted by a number of schools, the issue remained that changing the perceived role of the school involved more than a policy shift or an education decree. Even as far back as 1900, according to Diane Ravitch,

> *A journalist from one of the leading progressive magazines observed several New York public schools. She found some schools where modern methods were followed, but others where teaching was harsh and repressive. In one school, "I did not once hear any child express a thought in his own words. Attention was perfect. No pupil could escape from any grade without knowing the questions and answers of that grade."*[A11]

Teachers were not only unfamiliar with the new focus of learning, but they had few tools to enable the shift. Creating authentic learning experiences while at the same time bearing the responsibility to transmit knowledge, the role for which teachers had been trained, to a class of 30 or 40 students was a challenge few teachers could meet. Their training and experience didn't prepare them for the new demands made by progressive educational leaders – whether in 1900, 1920, or 1970.

New Roles – What Has Changed

Until recently, the impact of technology on school has been disappointing, significantly less than predicted. Radio, television, movies --each one was envisioned as the next big thing--the way to deliver more learning to more people. All delivered less than hoped for because none created a significant shift in what was already in place. School was the repository of knowledge and its role to impact society would be carried out through the delivery of content, knowledge, and cultural norms, according to a schedule prescribed by the school, the city, the state, the country, rather than based on the needs, interests, and curiosity of the learner.

School, along the way, has refined this system of delivery--*more* content, *more* frequent assessment of delivery, with narrower curriculum paths to follow, standardized across states and countries. In the name of convenience, accountability, and equality, everyone would receive the *same* content, delivered at the *same* pace, in order to achieve the *same* anticipated outcomes (meaning some would excel, some would fail, and the rest would populate the middle of the curve). School performed its function via delivery, and technology supported this role rather than transformed it.

Many students could not succeed along this limited, precise path and, instead, left school early, dropping out and joining the labor market. In the 19th and 20th centuries this wasn't too big an issue since there were many jobs to fill that depended on unskilled manual labor and the willingness to do repetitive work. According to Roger L. Geiger, Distinguished Professor of Education at Pennsylvania State University, in the early 1900's, "there was nothing to be done with a bachelor's degree that could not also be done without one."[A12]

Today, well into the 21st century, this is no longer the case. Repetitive, simple tasks can be done by machines, and even certain service workers are being replaced. Drop-outs today face a bleak future. The delivery system, instead of serving its societal role, is creating a bigger set of issues. The time is ripe for change.

Fortunately, as the old system is beginning to fail, a new one is developing, one based on many of the ideas of the progressive education leaders of the previous century, now made possible by technology. Not only has technology shifted directions, moving from mass to individual delivery, information now flows in two directions. Users are empowered to not just passively receive content but to personalize what is received and to create and share new knowledge and ideas with the world.

Technology has made it possible for individuals, rather than organizations, whether political or corporate, to be the creators of content and to use this content to build new ideas. Today's technology means every student everywhere can build a personalized learning journey, in which they are both the learner and the driver of learning. In the words of John Dewey, but now understood with new meaning, "Knowledge is no longer an immobile solid; it has been liquefied. It is actively moving in all the currents of society itself,"[A13] and, we can add today, in all directions.

Where does this leave School? No longer the deliverer of content, knowledge, culture, it now has the tools to provide learning environments envisioned by educators such as Rousseau and Dewey. With content everywhere, schools have lost their status as the repository of information, but does this mean they are without purpose? Or does freeing them from content delivery to more authentic and substantive learning, curiosity and passion driven, provide greater opportunities for education?

But, these opportunities for learning and education will not happen magically by just deploying laptops or tablets to students. As was seen in the open-class movement of the early 1970s, removing walls alone, without a deeper understanding of how to implement change, did not shift how learning took place. Without rethinking learning and teaching as well as the total learning landscape, neither classroom, school-wide, or systemic change will take place.

Part I - Three Essential Elements

Essential Element 1
The North Star: Vision, Mission, Goals

PRINCIPLE: The success of every initiative is dependent on how well its implementation is aligned with a shared vision, mission and goals that truly reflect the role of school in the new context of a digitally-rich world.

> "Our goal must be to find ways in which children can use technology as a constructive medium to do things that they could not do before;
>
> …..to do things at a level of complexity that was not previously accessible to children.
>
> — **Seymour Papert** [B1]

What should school be like in an age of abundance and exponential change?

Why start here?

Why not begin by purchasing devices, giving them to a teacher, and seeing what happens?

Why is this different from how schools have integrated other technologies, like projectors, etc?

Why start here?

Schools today are fortunate. They have the great benefit of hindsight. Over two decades of classroom and school 1:1 initiatives provide us with a vast range of experiences and more than enough outcomes and results to be able to see patterns and draw a number of clear conclusions.

Essential Element 1

Of course, all these experiences, stories, and successes (or failures) teach us valuable lessons only if we are willing to pay attention and learn from them. This framework is a result of examining what we have learned with the expectation that we will continue to be learning.

1:1 is not a new idea, it has withstood a good test of time. At this point in the 21st century, it is not a revolutionary idea to decide to supply each and every student with a personal, portable, fully functional device to be used anywhere and anytime. But 1:1 is more than this. It describes the relationship contemporary learners have with the technologically-rich world around them, and the changes that can be brought about because of the opportunities this technology makes possible for both learning and teaching can bring about a revolution.

My own philosophy is revolutionary rather than reformist in its concept of change. But the revolution I envision is of ideas, not of technology.

It consists of new understandings of specific subject domains and in new understandings of the process of learning itself. It consists of a new and much more ambitious setting of the sights of educational aspiration.

Seymour Papert, *Mindstorms* [B2]

One very clear lesson learned is that purchasing and deploying devices without knowing how they will be used is essentially, "the tail wagging the dog." In this era of funding struggles for education it is also a waste of money.

Knowing what you want to do and where you want to go from the start helps you build the roadmap you'll use throughout your initiative, from planning and design, through implementation and ongoing growth. That is why so much of this book is focused on the importance of a clear vision and actionable goals that can be well communicated to, and understood by, all those affected by the initiative, from teachers to students, to parents and the community as a whole.

And that is why we start here, first by understanding the current social, political, economic, and digital context and what it means to young people today and then by creating the vision and goals you'll need to guide you each step of the way.

If you don't know where you're going, you might not get there.

Yogi Berra

Yet when it comes to talking about defining a vision for our schools, we too often seem to stumble. Most schools today have developed a vision statement which many hang in the front lobby of their school - but what purpose does it serve? Does it genuinely guide decisions around the learning experiences of the young people in those schools or is it simply a statement of reassurance or an endorsement of legacy practice designed to comfort parents?

Essential Element 1

In too many schools it is the latter, and, as a consequence, it becomes more a school badge than a guiding light, more a motto than a statement of intent, and more a description of what is currently happening than an inspiration for what could and should happen.

Vision statements that provoke bold and ambitious outcomes for their students cannot be made on promise alone; nor can they be speculation. First and foremost, they must be built on a deep understanding of the context in which our young people are growing up and the compelling case for change that now demands.

What is the compelling case for change?

> …..the rise of a new Generation P, for 'participatory' (Gee 2003; 2004; 2005; Jenkins 2006a; 2006b). Here we will just consider the example of young people living in new media environments.
>
> Not simply vicarious viewers of movies, they play computer games in which they are the central character and in which their actions and identities in part determine narrative outcomes.
>
> Not simply listeners to the top forty songs on a radio station's play list, they create their own playlists on their personal listening devices.
>
> Not simply consumers of broadcast television, they choose amongst thousands of television channels and millions of YouTube clips; they even choose their own viewing angles on interactive TV or make their own television programs and upload them to the web.
>
> Not simply readers, today's literacy experiences as often as not also position readers at the same time as writers—in wikis, or blogs or their Facebook and MySpace pages, or small messaging spaces such as SMS or Twitter.
>
> Not simply consumers of pre-packed products, they become 'prosumers' of products which allow customization and even consumer contribution to the shape of the product for other consumers.
>
> Traditional relationships of knowledge and culture are profoundly disrupted, and even the terms of the either/or differentiations we have hitherto used to describe these relationships: creator/audience, producer/consumer, writer/reader.
>
> The key to these changes is an intensified cognitive and practical input on the part of previously more passive recipients of culture and knowledge, a shift in the fundamental direction of the flows of knowledge and culture, a transformation in the balance of creative and epistemic agency.
>
> — **Mary Kalantzis and Bill Cope** [B56]

Essential Element 1

Learners today live in a world very different from the one in which many, if not all, of their teachers and parents grew up and went to school. Although we often hear that these young people don't know what a world without technology is, stop and think what that means.

They are used to a worldwide, reciprocal connectedness to news, events, and people. Although connectivity to the outside world was available to previous generations, it had mainly been one-way and via limited media, with today's adults, in general, the recipients, not initiators, of the information and entertainment. Even those students today with the least amount of connectivity know about Instagram, Twitter, and Facebook. While many older people have been thrilled to be able to suddenly connect with family and new and old friends via Facebook, young people in the developed world are growing up in a world where this is natural and the norm. Not thrilling, but expected. And when this type of connectivity isn't available, the situation feels unnatural.

This pervasive connectivity influences more than just entertainment and social ties. It has created a new relationship between individuals, ideas, and the collective construction of knowledge.

Every industry and every organization will have to transform itself in the next few years, in multiple ways, or fade away.

Tim O'Reilly, *What's the Future of Work*[57]

As you complete the steps in **Essential Element 1,** you are laying down the foundation for your initiative. Do not skip this element or underestimate the importance of these actions.

By the end of the actions for this element, you will be ready to prepare a document outlining the reasons for your initiative, your vision, and your goals. This document will not only guide the work done as your initiative takes shape, but will be the heart of your communication plan.

Essential Element 1

Before Continuing

Your first action should be to assemble a small planning committee (in some circles this is called the 'guiding coalition') that includes the following people as well as a Project Team Leader who will be responsible for keeping the planning and initiative on track. You will expand this team once you get to **Implementation**.

Project Team Members

- District and/or school leaders, including principals
- Director of Curriculum and Instruction
- CFO
- CTO
- Director of Communication
- Project Leader

You may also want to include a parent representative.

Project Team Leader

Your school's superintendent or visionary should select the project leader. The person chosen for the job needs to have respect, influence, and passion for both project administration and instruction. Ideally, the project leader should also have experience working with parents and community leaders. If necessary, consider appointing two people to lead jointly.

Essential Element 1

Understanding the Context

Imagine a world in which every single person is given access to the sum of all human knowledge.

Jimmy Wales, Wikipedia Foundation [B55]

Students are interacting with the world and others in a variety of new ways, which is changing the role and expectations of school.

To grasp what this means, begin by exploring three key areas:

- The dramatic shift to digitalization and the impact that is having on our lives.
- How today's learner interacts with the world.
- The impact of ubiquitous access on learning

Much of the conventional advice offered to workers and to students who are preparing to enter the workforce is likely to be ineffective. The unfortunate reality is that a great many people will do everything right – at least in terms of pursuing higher education and acquiring skills – and yet will fail to find a solid foothold in the new economy.

Nonetheless, the idea that technology might someday truly transform the job market and ultimately demand fundamental changes to both our economic system and the social contract remains either completely unacknowledged or at the very fringes of public discourse.

Martin Ford, *Rise of the Robots* [B58]

Essential Element 1

Critical Conversation: Digitalization and the 21st Century Child

Children have always learned who they are and where they come from by first observing, living with, and interacting with their parents and their family. A young child's world tends to be very local – the home, the street, the neighborhood. The impact of outside influences whether from television, radio, or movies, while having some effect, tends to be minimized by the lack of interaction each of these requires. A child may react to a television program, but the television doesn't respond; there is no conversation or joint activity or other way for the child to feel this is an inherent element of her personal culture.

But today, even babies are interacting with technology in new ways. No longer passive watchers, they can touch a screen and see a reaction. They are choosing, whether intentionally or not, the direction in which the program or app or story goes; their actions produce some response. As they get a bit older, their relationship to technology colors their interactions with other media. The videos of children trying to move or interact with print media as if it were responsive to touch highlights that for even very young children (some videos are of children as young as one-year-old) their experiences with interactive media change how they interact with the world. How does this change their expectations as they enter school? Are they the same as students 100, 50, or even 30 years ago?

Numerous educators, thought leaders, and futurists write about the current generation, born in a world where technology is ubiquitous, where communication is cheap and instantaneous, where information is always available. They are always connected to their friends, to the media, to the world at large in a way that would have seemed like science fiction only a few decades ago. And while everyone then thought the science fiction future meant spaceships, flying cars, and astronaut clothing, what has really shifted is our ability to connect and communicate instantaneously, making everything in the now, as opposed to having been sent from the past. They perceive time differently, expanding their awareness of the present while shrinking their sense of how long the present is. Anything more than a few hours old on Twitter, for example, may not show up on their visible Twitter feed, making it seem old. The present is no longer this year, this month, this week – it's this minute.

The people with whom they connect and who help them shape their thinking, are from all corners of the planet and their opportunity for influence and impact extends far beyond their immediate location.

What is more, in this current state, young people have the opportunity to shape how they interact with the world. And although their actions are not without mistakes, the sense of empowerment is heady. It is one that provides immediate responses to their questions. One in which hereto unheard of amounts of information and content are readily available, usually in the palm of their hands. They literally have the world at their fingertips - until we take that away, shut it down, limit its availability.

Essential Element 1

This is the world today's children understand. This is the one to which they belong.

This shift isn't just anecdotal. Studies are showing that people are changing how they remember information when they know they have the possibility of accessing that information later. The internet has become an extension of people's minds and memory.

Searching for the Google Effect on People's Memory [B3]

... the interplay between humans and computers will change the behavior of both. The intriguing question is: Over the years, how will the thinking of the learners (students) evolve?

Evolving the Definition of Computational Thinking [B4]

Many people are familiar with Marc Prensky's trite interpretation of digital natives and digital immigrants but this is about a deeper difference, one that focuses on the impact of their understanding of time, space, and agency. Our children inhabit a digital world, one in which time and space are evaluated differently.

Those of us born before this digital era, having earlier experienced a different culture, see the world differently, even if we are very computer savvy. We have experienced the pre-technology world and bring a different contextual framework to our interactions. Our sense of time and space, availability and scarcity, openness and privacy, was shaped in a pre-digital age. This isn't a question of how profound the use of technology is by one group or the other or even which tools they use, but rather how our earliest experiences shape our interactions and understanding of the world.

How students see themselves is the starting point for learning.

More narrowly (the illusion goes), how they see themselves as learners is increasingly up for simple reconfiguration. What you post, who you tag, your avatar, you emojis, spelling, syntax, all digital expressions of self. The non-social internet is gone; social transactions are the single greatest currency of connected, digital spaces. It's not purely social, nor is it merely media. It's certainly more than commerce or media consumption, but it's strangely none of this. It's the careful packaging of consumable spectacle.....

... Identity is the learner insofar as they see themselves and are seen by others.

When Students See Themselves as Digital [B5]

Essential Element 1

Young people's references and sense of time, space and agency today are not mere modifications of the analog world; rather, they are shaped in a completely different dimension. School, on the other hand, is an analog world. Young people are expected to leave their way of thinking at the door.

Some educators ignore this or feel that it's important for students to experience the analog world. Either out of choice or ignorance, they do not take into consideration the student's culture or point of reference. It's no wonder that so many students do not engage with school, and, in the extreme, drop-out.[B6]

The cultural disconnect is huge. A key role of schools is to pass along the culture of a country, which though built from past experiences, exists in the present. Are current schools reflective of today's culture?

How have schools reacted? Most have infused technology into an analog world while expanding the implementation of the old culture by adding more of it, more content, more quality assurance testing, more accountability - using technology to amplify what they were already doing. Few have tried to rethink the school from the perspective of the contemporary learner.

Whereas the world around school has changed because of the rapid development of new mobile technologies, always on, always around, school has been particularly resistant. Massive, conservative, each small change must spread throughout the system before another change occurs.

Not all children are born immersed in the digital culture, but all will need to succeed in a world immersed in technology.

> *Schooling alienates two-thirds of kindergarten students by the time they reach ninth grade.*
>
> **Michael Fullan,** *The Principal: Three Keys to Maximizing Impact*[B59]

Essential Element 1

Questions & Actions

As you rethink learning for the 21st century, you must first understand how the world is changing and the global conditions driving the need for universal digital access. Schools or districts are moving to 1:1 because they recognize the new global economy is dependent on new skills, perspectives and knowledge-based work. Students need to be totally prepared for the technology-rich world in which they live and learn.

1. Read and discuss the ideas in **Learning from the Extremes**.[B7]

2. Begin to answer the following questions: What are the realities of the current technology-immersed environment in which today's young people are growing up? Within the context of the current global economic climate, what are the implications for schools?

Essential Element 1

> *The transformation of work requires much more than a mastery of a fixed curriculum inherited from past centuries.*
>
> *Success in the slowly changing worlds of past centuries came from being able to do well what you were taught to do.*
>
> *Success in the rapidly changing world of the future depends on being able to do well what you were not taught to do.*
>
> — **Seymour Papert and Gaston Caperton**[B8]

Next, consider the impact these changes are having on how today's learner interacts with the world. Schools and districts are moving to 1:1 because they recognize and acknowledge that the world outside of school for today's young people is awash in technology as many young people interact with and use powerful technology on a daily basis. Young people expect to have access to rich content and collaborate frequently and easily.

1. Begin by answering the following key question:

   What are your expectations around what technology makes possible for students?

Essential Element 1

2. Once you've discussed the question above, consider this question:

What expectations do your students have?

Compare and contrast your answers to both questions.

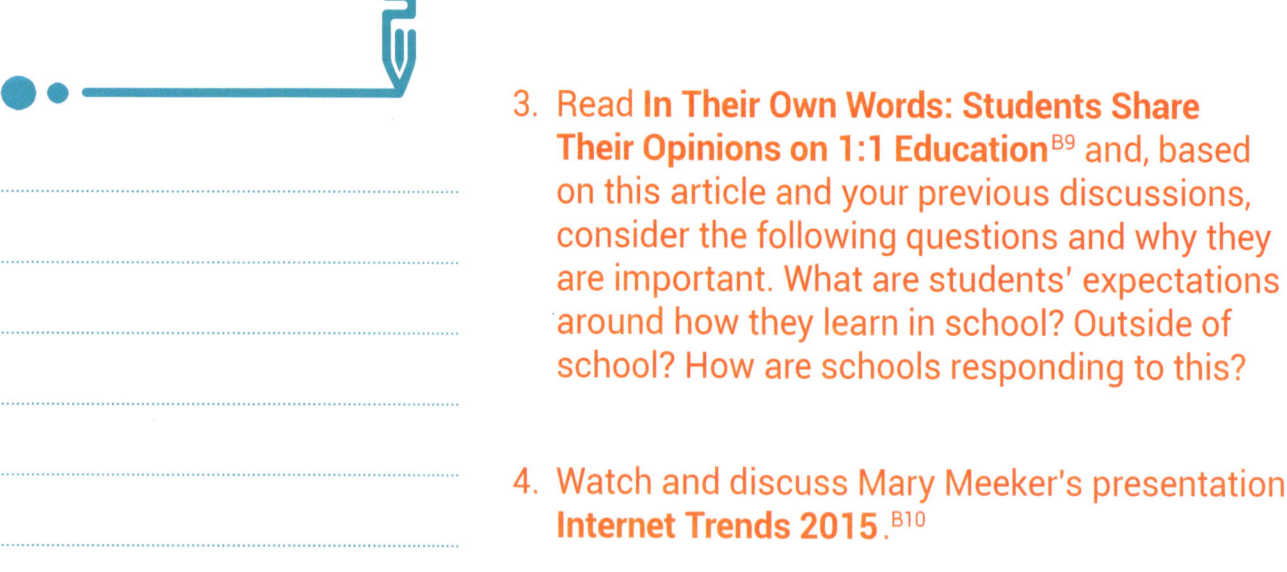

Image Credit: George Couros, used with permission

3. Read **In Their Own Words: Students Share Their Opinions on 1:1 Education**[B9] and, based on this article and your previous discussions, consider the following questions and why they are important. What are students' expectations around how they learn in school? Outside of school? How are schools responding to this?

4. Watch and discuss Mary Meeker's presentation **Internet Trends 2015**.[B10]

Essential Element 1

> Finally, but of equal importance, you must understand the research around 1:1, the impact that is having on learning, and how to best prepare young people to become contributors to this future society and economy. Schools or districts are moving to providing universal digital access because they recognize that with increased access to technology, there Is the potential to shift creative control from the select few to the many (sometimes referred to as the democratization of creative expression and innovation).
>
> What are some of the reasons an increasing number of schools are moving to 1:1?

1. Do research on 1:1 learning in schools in order to develop a better understanding of the benefits of and issues around implementing 1:1. Use this information to build a strong foundation for your initiative.

    Begin your research here:
    - **Benefits and Challenges of Using Laptops in Primary and Secondary School**.[B11]
    - **Powerful Tools for Schooling: Second Year Study of the Laptop Program**.[B12]
    - **The Impact of Maine's One-to-One Laptop Program on Middle School Teachers and Students**.[B13]
    - **Reflections on the 25th Anniversary of 1:1 with Steve Costa, Methodist Ladies College**.[B61]

2. Have everyone in your planning team watch the video **The Compelling Case for Change: What Is Possible**.[B14]

    This video focuses on two sites, a school and a district, both long-term 1:1 initiatives that exemplify how 1:1 can be used to transform learning and teaching practice. Discuss with your team what learning and teaching look like in these schools. Note the shifts in student learning, teaching practice, and learning environment you observe.

3. You may want to review and discuss the **4Up - Year 4**[B15] vodcast from New South Wales, Australia, that describes the development and impact of their laptop program, part of Australia's Digital Education Revolution initiative.

4. Discuss this statement:

    "Today we have schooling in an age of ambiguity and uncertainty."

    What do you think this means? Do you believe it's true? If yes, what impact does this have or should it have on what and how students are learning?

5. Read **Innovative Teaching and Learning (ITL) Research**[B16] and **The Case for Computing**.[B17]

Essential Element 1

Modern Learning Environments

> *The activity we call "learning" is walking away from the institution we call "school." Schools must transform themselves with new models of instruction that reach out into the world; must fashion new understandings of what powerful learning relationships look like between adults and students, and students and the world; and most importantly, must hone and develop skills that help students know when they know.*
>
> — **Kirsten Olson,** *Wounded by School* [B18]

Contemporary pedagogical insight comes from a better understanding of the realities of the modern learner's world and how they learn.

In addition, the intersection of three key factors is creating conditions for new learning opportunities. These factors are:

- Rapidly Emerging Web Technologies
- Universal Access
- Contemporary Pedagogical Insight

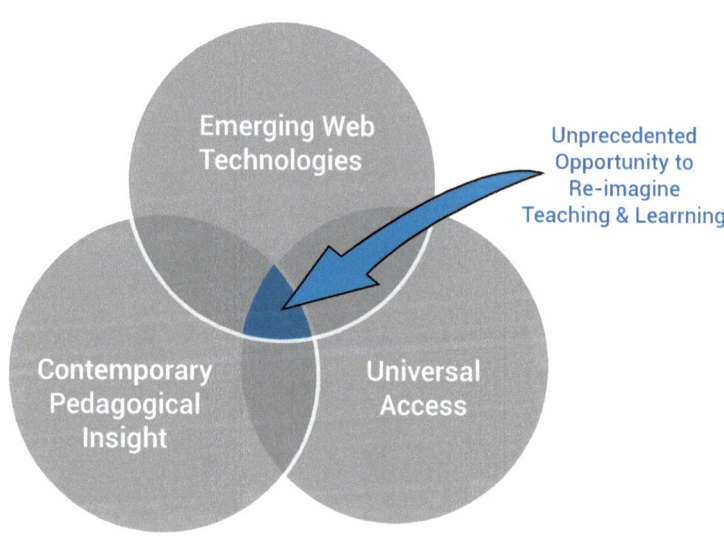

Specifically:

- **Where students are learning is changing.**
- **When students are learning is changing.**
- **What students are learning is changing.**
- **How students are learning is changing.**

Essential Element 1

Questions & Actions

How is technology contributing to these shifts?

Reflect on the four statements above. Begin to research and collect evidence of each of these shifts and consider what this means in terms of the role of school. Keep a record of these discussions, since the ideas developed as you answer the following questions will help guide later work.

Emerging web technologies provide an "architecture for participation." [B60]

These technologies are:
- Challenging traditional approaches to how our students learn.
- Challenging our assumptions about classrooms and teaching.
- Challenging our assumptions about knowledge, information and literacy.

What are the implications for our schools?

Participatory Culture Characteristics

- Shows relatively low barriers to artistic expression and civic engagement
- Has strong support for creating and sharing one's creations with others
- With some type of informal mentorship whereby what is known by the most experienced is passed along to novices
- Where members believe that their contributions matter
- Where members feel some degree of social connection with one another

"Not every member must contribute, but all must believe they are free to contribute when ready, and that what they contribute will be appropriately valued." [B19]

Henry Jenkins

Essential Element 1

1. Read and discuss **Confronting the Challenges of Participatory Culture.**[B19]

   What are the implications of this for schools? What is the meaning, importance, and impact of this? What opportunities do these make possible for learning?
   How is this changing our students' expectations about learning?

2. Roy Amara of the Institute for the Future has said, "We tend to overestimate the effect of technology in the short-run, and underestimate the effect of technology in the long-run."[B20]

   Discuss this statement with your team. Do we overestimate or underestimate the effect of technology on learning and school? How are these perceptions impacting expectations? Why do we appear to have such low expectations for what digital richness now makes possible?

> How well do existing pedagogies serve the needs of our young modern learners? Contemporary pedagogical insight comes from a better understanding of the realities of the modern learner's world and how they learn.

1. What are the prerequisites for a modern learning environment? Discuss this as a group and begin to create a list of what needs to be in place to start the process of building an environment for today's learners.

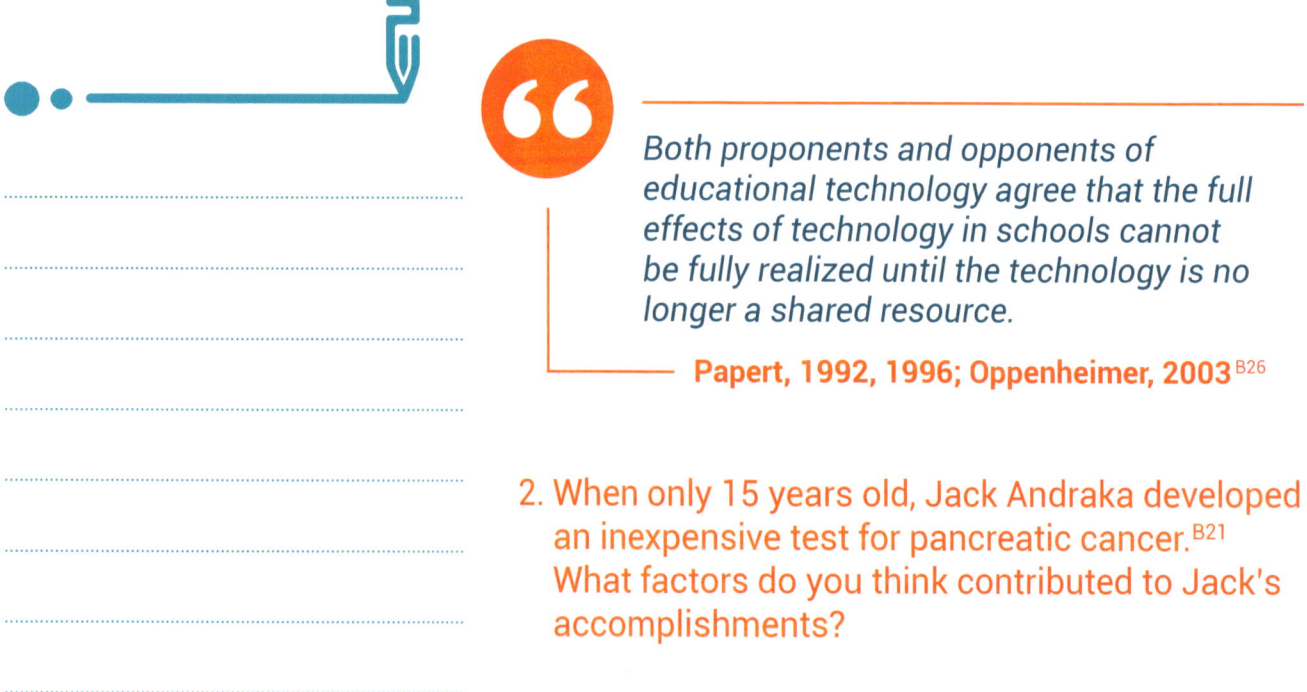

> Both proponents and opponents of educational technology agree that the full effects of technology in schools cannot be fully realized until the technology is no longer a shared resource.
>
> — **Papert, 1992, 1996; Oppenheimer, 2003**[B26]

2. When only 15 years old, Jack Andraka developed an inexpensive test for pancreatic cancer.[B21]
   What factors do you think contributed to Jack's accomplishments?

Essential Element 1

PLANNING OUTCOMES

As a result of your research, readings and reflections…..

What are the *key ideas* that are making the need for transformation so critical?

Compile your background research and record the key points in your plan.

You may want to begin by including:

Project Leader is
Program Manager is
Planning Team Members are
Background Research
Global conditions indicate:
Students today:
Research on 1:1 shows:

Essential Element 1

(Re)Define Your Vision

> The first thing all leaders must do, regardless of whether they are leading General Motors or IBM or a school system, is to clearly articulate a vision. Whether the leader calls it a strategic intent, mission, conviction, set of beliefs, or vision, it must be communicated clearly, compellingly, forcibly, and simply. And it isn't enough to talk generally about strategic goals or objectives. A vision must be communicated ceaselessly, indefatigably, and endlessly in all sorts of ways.
>
> **Warren Bennis** [B27]

Essential Element 1

What is a vision statement? And how important is it?

To understand its importance, it's best to start by understanding what vision is.

A vision statement describes how the world will look after you finish changing it or fully implement your mission. It describes where you are going.

Consider these vision statements:

Oxfam: A just world without poverty.

Habitat for Humanity: A world where everyone has a decent place to live.

Kiva: We envision a world where all people – even in the most remote areas of the globe – hold the power to create opportunity for themselves and others.

ASPCA: That the United States is a humane community in which all animals are treated with respect and kindness.

Can you picture the world to which each one is aspiring? Do you see where they are heading?

Crafting a vision is not a simple task. Each of the above statements, so simple, yet so elegant, probably took much discussion and went through many iterations. What they have in common is that they each focus on the future towards which all plans and actions are to be focused and against which all outcomes are to be measured. It is impossible to determine if you are moving closer to where you want to be if you have not clearly defined where or what that is. Each of these visions does just that.

Essential Element 1

Critical Conversation: Vision and the Art of the Possible

Does vision really matter? For many the answer is a given; to some, it is almost sacrilege to suggest that it does not. Yet where is the evidence of the impact it has in our schools today?

There are few schools in the developed world that have not spent a good deal of time developing a 'vision for their school', which is then usually found prominently in the entrance hall of their administration building.

More often than not, it looks something like this...

> Our vision at Imaginary School is to empower students to acquire, demonstrate, articulate and value knowledge and skills that will support them, as lifelong learners, to participate in and contribute to the global world and practise the core values of the school: respect, tolerance & inclusion, and excellence.

At best it is what you might call a 'motherhood' statement, or, rather, a vision that means everything to everyone. At worst, it says nothing. It was obviously drafted with the best of intent and is usually the result of many hours, sometimes weeks of meetings and/or workshops with 'community stakeholders', yet one must ask exactly how does it in any way have impact on the learning experiences of the students in that school?

The answer of course is that it does not, but rather what it does is serve as an affirmation of existing practice and conformity to the traditions of school as it is most familiar to us. None of this should surprise us, and, in fact, it should be expected, because it is the model of schooling with which we are all most familiar and the one through which every parent and teacher progressed in their younger years. So what we have is a self-perpetuating cycle, which in itself is largely responsible for the predominantly incremental shifts in our current model of school.

But are we satisfied with that? Do we really believe that a vision that sets the agenda for our schools should simply be an echo of the past or should it be looking further?

Beyond Magical Thinking - Finding A New Path

"What if we don't change at all ... and something magical just happens?"

Surely the accelerating rate of change we are seeing in the world around us demands that we must urgently rethink how we develop the visions we have for our schools. We can no longer accept visions that perpetuate a past model that no longer reflects the world today.

If we are serious about our vision being the guiding light for where our schools are heading, then we have to dig deeper, and challenge the assumptions on which our existing schools are built. But how does that start? Who is responsible

Essential Element 1

for leading the shift in thinking that is now so urgently required, and how will that ultimately impact what we do in our schools?

The first step is to initiate and lead a much wider, more informed conversation. The second is to understand the consequences and implications that arise from that and then develop a vision for schooling that truly reflects the needs of young learners growing up in the modern world.

We have to accept that many people are oblivious to the rate and scale of change, while at the same time the education sector has to date played such an insignificant role in leading a wider conversation around the impact this will have on our lives. If books such as **Rise of the Robots** [B28] and Mary Meeker's **Internet Trends 2015** [B10] report are not a wakeup call, then something is truly wrong. The recent *Techcrunch* article about disintermediation by Tom Goodwin, **The Battle Is For the Customer Interface**, is further evidence, where he observes:

> *Uber, the world's largest taxi company, owns no vehicles.*
>
> *Facebook, the world's most popular media owner, creates no content.*
>
> *Alibaba, the most valuable retailer, has no inventory*
>
> *...and Airbnb, the world's largest accommodation provider, owns no real estate.*
>
> *Something interesting is happening.* [B29]

Maybe we should not be puzzled because for the many people who still rely on traditional media, the paucity of serious journalism now found there too often appears to leave them bereft of learning anything worthwhile about the changes in the world around them. Few might have predicted that our newspapers and television shows would have sunk to the level of triviality that now dominates. If we include popular magazines and reality shows, at times all seems totally lost.

So for the most part, keeping well informed has largely been left to social media and the new digital forms that are taking the lead in helping people keep abreast of the shifts around us with emerging publications, such as **The Information** [B30] and **Quartz**; [B31] all of this, of course, supplemented by increasing numbers of books that are providing depth and insight for those who know they need to know more.

We have talked for years about a digital divide and as predicted this is now manifesting into a deep information divide. The shifts in the media through which we access information and our ability to effectively use them is creating a small but well informed elite, while the vast majority of people seem too often comfortably distracted by trivia, and blissfully ignorant of the potential impact that change is having on their lives.

So what are we doing about it? Why is there not more public dialogue led by educational leaders around these unprecedented shifts in our lives, and, more importantly, where is the thought leadership around how our schools can best adapt to the realities of the world that

Essential Element 1

their students will be heading into in the 2020's? Do we really still think that the order of magnitude of the changes that we are now seeing can, in any way, be accommodated by incremental shifts in our existing school structures?

However, while we may be bothered and puzzled, we should also be optimistic because I also sense that we are starting to see a shift in thinking. Having recently completed a series of workshops with school leaders across four countries, I sense a growing acknowledgment from them that more fundamental change is now urgently required. So the real challenge is what might that look like, and just how do we make it happen?

For one, I think to date we have been going down the wrong path. Too often we have talked about transformation as if it was a recipe that if followed gave us a 'right' answer. Whether it was through the 'magic' of introducing technology or the 21st Century or Deeper Learning touchstones, we have been refining rather than reimagining, tweaking rather than transforming.

Maybe it is time for a very different approach which should start by looking outward, rather than inward. We need to seek input and insight from beyond our school walls and have a much more open and broader conversation around the sorts of schools that might be best for our young modern learners. The dialogue to date has too often been framed very narrowly by an academic elite supported by high profile writers and 'experts' who too often fail to engage school leaders, let alone the broader community, policy makers, and, most importantly, parents. There has been too much speaking to the converted and shouting into an echo chamber, when what we should be focusing on is a wider dialogue with a broad cross-section of the community who we can then bring along with us, rather than impose our ideas on them.

While such a strategy might be met by some with suspicion and negativism, I would contend that ultimately it gives us an opportunity to build a strong collective of stakeholders from across school communities who can become advocates for transformation, rather than poorly informed opponents.

My experience is that when you talk with most people about the skills and competencies they think our young people will need for their future, they are in fact often in line with much of the more radical thinking of where school transformation should be heading; but they nearly always 'revert to norm' when asked what that might look like as they have little or no experience of what other options might be more preferable. This is where we have to step up.

Who's Leading the Conversation?

I think it is time that school leaders take a more proactive, even provocative role in engaging staff, parents, and the wider community in debate around the transformation our schools must undertake. This should include media and politicians and must rise well above the legacy ramblings of the loud minority who want to cling to traditional practice like it is a life raft. However, despite that, I truly believe that if we can also invest time to help them be better informed about the rate of change in the world around them, then they too will, by and large, be supportive.

Essential Element 1

As we head into the '20's, we must raise awareness of the challenge schools face to a level where the wider public have real awareness and ownership of the need for transformation of our schools, and subsequently we have to broaden the scope for what that might look like.

If our vision for schools continues to be built on simply improving traditional literacy and numeracy then frankly we are doomed. If that vision embraces the new literacies of a highly connected world,[B32] and mathematical thinking along the lines of what people such as Conrad Wolfram[B33] are articulating, we are on the right track. But, again, how can our parents and the wider community truly understand these possibilities if it is not coming from educational leadership?

Again, we must respect the fact that none of our parents were educated in technology-rich classes. None of our parents have any firsthand experience of the potential and possibilities that ubiquitous access provides for their kids; and if we continue to fail to engage them in a meaningful dialogue about what they might be, then none of our parents can be advocates for the shift we now know is so necessary.

The dialogue around school reform has to date largely been a call for school improvement, while the concept of transformation seems have been lost in the search for identity, a brand, a recipe or a formula. We need to get it back on track. It's about moving beyond better, and thinking different.

This is now a time for a new generation of school leaders who can take back the public dialogue from journalists and politicians and build a strong, broad base of support for modern schooling, for schools that truly reflect the context of the world around them and create very different places of learning that grow with, not in spite of, the support of the wider community.

Implications: A Conversation Becomes a Vision

Such a process provides a platform on which a vision for contemporary schooling can now be developed, and here is where the real challenge and opportunities lie.

> *Vision is the art of seeing what is invisible to others.*
> Jonathan Swift

What comes next is what makes contemporary educational leadership. It is about how a leader now takes his or her understanding of the context in which our students are growing up and is able to explore the implications for what that means for teaching and learning. This requires bold and ambitious thinking; it requires creativity and imagination, and, above all, it requires a willingness to challenge many of the accepted norms of traditional classroom practice.

This is about what we teach and how we teach. It is unsettling and uncomfortable, but necessary. It is about developing a vision that *truly* reflects the choices the school community has made about what really matters and what is worth doing.

Essential Element 1

The reality of the past meant that for many decades, educators pre-determined the knowledge they judged as being worthwhile. This predetermination was a compromise. Some stuff was put in, other stuff left out, but in the pre-digital world there was little choice. You had around 1,000 hours per year to do your best to 'cover' the curriculum you were given. Curriculum writers around the world would do their best to guess what knowledge would be of most benefit to their students, and they designed the curriculum accordingly.

This thinking is obsolete; it has no place in the technology-rich lives of our students who are today challenging the very foundations on which our schools have been built. Our kids are looking for relevance, authenticity, and collegiality embedded in everything they learn, and if it isn't offered within their school, they simply look elsewhere.

Modern leaders think of their curriculum as a means to an end, rather than the end itself. It's about the questions that matter to young people, not the answers that don't, and it's about essential or powerful ideas, rather than the distracting focus of high-stakes tests. In that narrative, curriculum is selected to meet students' needs at the moment, not delivered in a "just in case" way whether they need it or not.

This then is now made possible in a digitally-rich, highly connected world which in turn requires a math teacher to enable her students to 'think mathematically', not complete tedious 'hand-calculated arithmetic'. It says to a history teacher, "don't teach your students about history, but rather allow them to be historians." And it says to a science teacher that his goal is to imbue his students with a passion for scientific discovery, rather than simply retell the science that is already known.

Above all, this is a leadership attribute about lifting expectations of both teachers and students because it takes away the security blanket of a content focused curriculum and replaces it with the rigor of one that is built on empowering strategic outcomes.

No longer will the modern educational leader see traditional rote knowledge as acceptable, nor will she believe that the limited expectations we had of our pre-digital curriculum are in any way appropriate today.

As we move forward, young people will only benefit from a view of curriculum that is agile and iterative. It is a perspective that sees curriculum not weighed down by insecurity, but rather inspired by possibility.

Will Richardson, *Eight New Attributes of Modern Educational Leaders*[B34]

Essential Element 1

So what does all that mean for a school's vision?

It means the world has changed, and so must our schools, and the only way we can lead that change is if we understand context, articulate the possibilities, and support the imagination and creativity that is required for us to stay pedagogically relevant. It is no longer appropriate to be delivering content as if we were the sole source of knowledge. It is no longer suitable for us to define a scope for learning that is contained within a classroom or school walls.

Today our vision has to be about setting a course for a journey of discovering new ways of thinking, new ways of learning, and new pedagogies that reflect that. It is about seeing what is invisible to so many, and, more importantly, about helping them reach what you see.

It's about the art of the possible; what this digitally-rich, highly-connected world now makes possible, and to lead that, our vision is defined not by what we know, but rather by what our students will discover.

So vision does matter. In fact, it might be the only thing that does matter, for if we get it wrong, our schools will fade into irrelevant oblivion. If we get it right, it will set a course for what modern schools could and should be.

Bruce Dixon

Consider the following school and district vision statements. Do they create a vision of how their schools will look and what they will be in a few years?

Brightworks School, San Francisco, CA, USA

It started with a question: what if school could be the most interesting place in a child's life. We thought we should try to make that place: **a learning community driven by kids' passions and interests, where their curiosity fuels the curriculum and where their work really matters.**[B35]

Austin Independent School District, Austin , TX, USA

Every AISD student is a designer of the changing world. They are producers and contributors, not just consumers. Engaged, not compliant. Persistent learners and doers with flexible skillsets that help them thrive in a world that is connected and in perpetual change. AISD is a community of learners connected to other communities of learners.[B36]

New Tech High Vision

Students learn in an innovative and professional environment fostered by the use of advanced learning methods and technology. Both staff and students understand the commitment necessary to implement a rigorous and relevant curriculum, one in which technology, standards, and skill development are embedded.[B37]

Essential Element 1

Princes Hill Primary School, Australia

School visions may change over time. A first order of business for Principal Esme Capp of Princes Hill Primary School was to develop an "ever-evolving shared vision for the school."

The school's current vision statements
- *Children are active, important members of a variety of communities, for example: family, school, ethnic cultures, multimedia and friendship groups – their understanding of the world develops through these social and cultural interactions.*
- *We learn through active participation, using the many forms of expression.*
- *We learn through critical engagement in complex, purposeful contexts where relevant connections are made to our world.*
- *We learn through consciousness of thought where we re-configure pre-existing understandings and concepts.*
- *We develop motives to learn through positioning ourselves within social situations.*
- *We learn through the unity of emotions and intellect.*[B38]

Today's learning environment has come about because of the convergence of emerging Web technologies (the world in which our students live), contemporary pedagogical insight (how students learn), and universal access (1:1). But, essential as these three conditions are, they alone are not enough to create change. What pulls these together and truly shapes an initiative is a clearly articulated vision that will guide all planning, design, and decision-making.

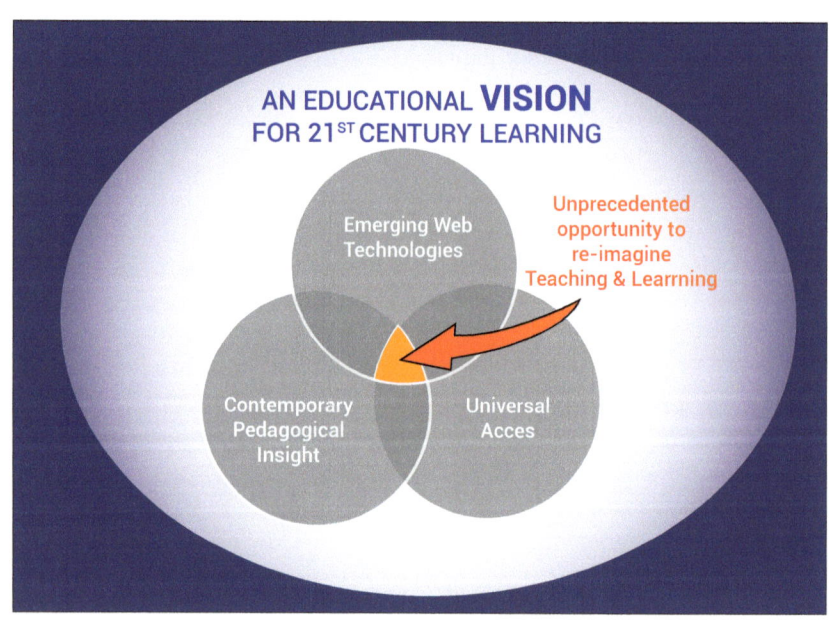

As you define your vision, consider these questions:

- **What is the vision you have for the future of your schools?**

- **What must learning at your schools be like in 5-7 years' time?**

- **How would you articulate a vision that provides clear direction for the school for the next 5-7 years?**

- **How will access to ubiquitous technology shift what learning in your schools looks like?**

Essential Element 1

Questions & Actions

> Your vision statement should serve as the foundation of your decisions and inform the direction for all those involved in the initiative's implementation. What is the vision you have for the future of your schools?

Have each team member read at least two of the following and discuss the ideas as a group:
- A Policy Agenda for a 21st Century Education [A1]
- Vision for Education: The Caperton-Papert Platform [B8]
- Why School? How Education Must Change When Learning and Information Are Everywhere [B39]
- Learning in the Digital Age [B40]
- Eight Forces for Leaders of Change [B41]
- Enabling Transformation with Strategic Planning, Organizational Capacity, and Sustainability [B42]

Essential Element 1

Your vision statement must articulate how your 1:1 initiative will transform the learning environment and improve student learning outcomes. How would you articulate a vision that provides clear direction for the school for the next five to seven years?

1. Watch the workshop video **A Vision for Learning, an interview with Esme Capp**.[B43]

   In this video, Esme Capp, principal at Princes Hill Primary School, Australia, describes not only her school's vision but also how the school has defined goals and outcomes based on this vision. Consider how she has developed these goals based on her school's vision of how children learn. (You may also want to read an expanded version of the interview.[B44]) Outline how you could have similar conversations at your school.

2. You may also want to watch the workshop video **Learn From the Leaders**.[B45]

   In this video, seven school and district leaders describe their visions for learning. How have these schools articulated a vision that can guide and help define all aspects of learning and teaching? Consider these examples as you define your vision.

Your vision must clearly scope the potential technology-richness it offers learners and provide a platform and permission for teachers to take risks, to innovate, and to reimagine their teaching in the context of the modern, anytime, anywhere learner. How will access to ubiquitous technology shift what learning in your schools looks like? What role will theories of learning play?

Few people are really able to see future possibilities, more specifically, what a 1:1 environment might make possible, without some support to help them.

1. Watch and discuss this video interview on **Future Learning**.[B46]

   How will your vision support creating "relevant, worthwhile experiences for young people?"

Essential Element 1

2. Have each individual in your leadership team review the list of vision statements in the **Articulate Your Vision** activity in **Part III - Tools and Resources** and select the three statements that best articulate his or her vision. Feel free to add other vision statements that you feel may better articulate your vision, but be as precise as possible.

Once each individual has selected three statements, discuss these selections as a group and, as a group, narrow down your selections to the three statements, in order of priority, that best articulate the school/district vision. Your vision needs to paint a picture of the future and provide direction for your initiative.

> ....no generation has had more ability to create its own future.
>
> **Fred Deakin,**
> *Three key shifts in the design industry* [B47]

An important aspect to consider is what you think the challenges are that you might experience in building and sustaining a technology-rich 1:1 learning environment.

1. Write down the challenges you foresee. In your planning, make sure you address each of these and design strategies to overcome them.

Consider not only the factors that have contributed to positive outcomes for 1:1 initiatives, but also those that have led to 'failed' or discontinued initiatives.

There have been a number of well-publicized 'deployment failures.' Careful analysis of these will show that either the schools had skipped some key steps in planning and implementing their initiative, that their assumptions, expectations, and/or evaluations were not built around the opportunities ubiquitous technology makes possible, or both. It is important to look at these cases, too, and learn from them.

Essential Element 1

Why have so many ICT initiatives in the past had limited impact?

- Not policy initiatives but **projects—** policy in bits and pieces
- Policies are **not above politics** - current policies are replaced by the new government
- The policy underpinning the initiative focuses **only on ICT**
- The policy provides a **short-term strategy** without a sense of where this will go in the long-term
- ICT policies are too often based on **incorrect assumptions**, which create **unrealistic expectations** for what can be realistically achieved
- The policy is **organizationally isolated**
- The policy does **not specify measurable goals**
- Researchers describe a **gap** between **rhetoric** in government **policy** and **reality** of **education practice**

Intel, *Transformation Framework*

Essential Element 1

1. Review the list of reasons so many ICT initiatives have limited impact. As you go through this framework and build your initiative, review these points regularly to ensure you don't find yourself dealing with these issues. Include these in any evaluation processes you establish.

2. Read and discuss the following articles to examine why they did not succeed:
   - **Seeing No Progress, Some Schools Drop Laptops,**[B22]
   - **L.A. Unified takes back iPads as $1-billion plan hits hurdles,**[B23] or
   - **Why Hoboken is Throwing Away All of its Student Laptops.**[B24]

3. The effects of these troubled implementations may ripple out to impact other decisions around change, the use of technology, and funding, as can be seen here: **Science, math and art valued more than technology in education poll.**[B25]

   How did the problems and perceptions of the Los Angeles initiative influence these poll results?

One Last Note: Mission vs Vision

A mission statement outlines how you plan on acting to create change or fix a situation. It explains why you exist, describing what you do, for whom, and how you do it.

For example, who could forget the famous Star Trek mission:

> *These are the voyages of the starship Enterprise. Its five-year mission: to explore strange new worlds, to seek out new life and new civilizations, to boldly go where no man has gone before."*

Closer to home, here is New Tech High's mission statement:

> *Our mission is to inspire students to be responsible, resilient, and personally successful in the rapidly changing 21st century, and to be a student-centered model for educational innovation.*

Mission and vision, though different in focus, go hand-in-hand, each complementing and illuminating the other.

Mission cannot exist without vision. It's not enough to know what you do to get where you're going, you need to know where you're going, too.

Essential Element 1

PLANNING OUTCOMES

As a result of your research, readings, and reflections…..

Write your **final vision statement**. Define a clear, actionable vision that:

- **Provides the best opportunities for students as functioning members of the broader society.**
- **Represents your beliefs about the role of school in the 21st century.**
- **Is constructed as a voice for the whole school community.**

Our Vision Priorities

Top 3 statements are:

1.

2.

3.

Our Vision Statement

Essential Element 1

Clarify Goals and Policy Priorities

> Children as participants in the 21st century need to understand themselves as learners, learn to work collaboratively, engage in new technologies, learn how to access new skills and knowledge and develop the skills of thinking creatively, laterally and critically.
>
> The school's beliefs about learning, pedagogical practice, organisations, structures and the physical environment reflect these core principles.
>
> **Esme Capp, Principal, Princes Hill Primary School, Australia,** *From the Principal* [B48]

Create measurable, actionable goals and expected outcomes that help achieve your defined vision. Goals are the shorter term steps you will need to take to achieve your vision. They help you map out your actions, more precisely describe your vision, and clarify your mission.

Essential Element 1

Questions & Actions

How will you move from articulating your vision to clarifying and defining more specific goals?

There needs to be a shared understanding of what the goals are and how you plan to achieve them, as well as what achievement of the goals may look like at different school and grade levels.

1. Outline a process for defining your goals.

Essential Element 1

> **Five guiding questions for change**
>
> 1. What are you trying to do?
>   *(Clear priorities, specific, measurable goals.)*
>
> 2. How are you trying to do it?
>   *(Clear, practical plans which are used regularly and updated.)*
>
> 3. How, at any given moment, will you know whether you are on track?
>   *(Good, steady, close to real-time data on key indicators. Monitoring routines, such as stocktake meetings with all key stakeholders involved.)*
>
> 4. If you are not on track, what are you going to do about it?
>   *(Agreed actions, followed up, tested in practice and refined if necessary. Always try something. Never neglect a problem once identified.)*
>
> 5. Can we help?
>   *(Constant ambition, refusal to give up. Focus on the goals, no distractions. Maintaining the routines. Analysis and problem-solving where required. Bringing to bear lessons from elsewhere.)*
>
> Michael Barber, 2008[B49]

2. How will you ensure learning and teaching practices throughout your school or district are aligned with your mission and goals?

3. How will these goals and your expectations and policy priorities align to deliver on your vision?

The following illustrations show how a country-wide vision translated to specific goals for schools in Singapore.

Essential Element 1

Singapore's Masterplan for ICT in Education

The following example shows how a clearly defined national vision and goals lead to enabling the alignment of a school vision, mission, and goals.

Since 1997, Singapore has introduced three masterplans for ICT in education. The most recent, Masterplan 3 calls for leadership that focuses on:

1. **Strengthen competencies for self-directed, collaborative, learning. (SDL & CoL)**
2. **Tailor learning experiences according to the way that each student learns best.**
3. **Encourage students to go deeper and advance their learning.**
4. **Learn anywhere.**

It also outlined strategies to achieve the goals in Masterplan 3, for example:

Outcome Goal
Students develop competencies for self-directed and collaborative learning through the effective use of ICT as well as become discerning and responsible ICT users.

Enabler Goals
- **School leaders provide the direction and create the conditions to harness ICT for learning and teaching.**
- **Teachers have the capacity to plan and deliver ICT-enriched learning experiences for students to become self-directed and collaborative learners, as well as nurture students to become discerning and responsible ICT users.**
- **ICT infrastructure supports learning anytime, anywhere.**

Further to this, each of these goals was further defined to ensure a shared understanding (see images below).

Essential Element 1

Students will
- Negotiate and set common goals
- Interactively contribute own ideas clearly and consider other points of view objectively and maturely
- Ask questions to clarify and offer constructive feedback
- Reach consensus and take on different roles and tasks responsibly within the group to achieve group goals
- Reflect on group and individual learning processes

Collaborative Learning

- Effective Group Processes
- Individual and Group Accountability of Learning

Students will
- Work towards completing individual's assigned tasks as well as help group members achieve group goals
- Rely on each other for success

Images credit: The ICT Connection

Self-Directed Learning

- Ownership of Learning
- Extension of Own Learning
- Management and Monitoring of Own Learning

Students will
- Articulate learning gaps
- Set learning goals and identify learning tasks to achieve the goals

Students will
- Apply learning in new contexts
- Learn beyond the curriculum

Students will
- Explore alternatives and make sound decisions
- Formulate questions and generate own inquiries
- Plan and manage workload and time effectively and efficiently
- Reflect on their learning and use feedback to improve their schoolwork

Essential Element 1

1. Review **The ICT Connection: Harnessing ICT, Transforming Learners**.[B50]

    This presentation highlights how Singapore outlined a national vision and defined goals to achieve this vision. Analyze how Singapore's national goals were used to develop regional and school policies and classroom goals, ensuring systemic alignment.

2. Based on your vision, create a common language for all stakeholders to understand and discuss the journey to the vision. Use this language to define your goals and help all those involved align their work to your vision. Alignment is crucial.

How will you develop goals that will effectively measure the extent to which you are meeting your objectives and delivering on your vision? How will you know you're on track?

1. Determine milestones as a way for you to gauge how well you are doing in this process.

2. Break down the steps you will need to take to achieve your vision. Consider what is negotiable and what is non-negotiable.

3. Develop a reasonable timetable for implementing each goal and achieving your vision.

4. Define the outcomes you expect for each goal and include opportunities for review and adjustments if you are not on track.

5. You may want to create your goals using the SMART protocol to ensure goals are **S**pecific, **M**easurable, **A**chievable, **R**ealistic, and **T**imely.

Essential Element 1

 What are your national/district ICT policies? Do they support your 1:1 vision and goals?

1. Outline how your national/state or district policies are:
   - Based on a long-term vision and have clearly articulated goals.
   - Focused on learning and not only on ICT.
   - To be aligned over time and reflected across the work of all stakeholders.
   - ...and, yes, Beyond politics.

 Reviewing case studies from other districts may provide valuable lessons relevant to your initiative.

1. Discuss one or two of the case studies found on the **AnytimeAnywhereLearners** website.[B51]

Essential Element 1

Evaluate - Continuously!

How can you structure evaluation so that it's continuous and impactful? What will be the scope and parameters of your continuous evaluation? Who will ensure the outcomes impact the ongoing program?

Effective evaluation of your initiative and its implementation fidelity is the backbone of its success. Without any genuine commitment to such a process, the impact of all the elements can be marginal and poorly understood. It underpins the ongoing development of a program and ensures the best possible outcomes for students, staff, and the school.

Reflection and evaluation are ongoing processes that can be both formal and more informal. A formal evaluation process and schedule should be established at the start of any initiative and an evaluation team should be set up.

Smaller, less formal evaluations can be conducted at various other times, in order to confirm that initiative steps are aligned with your goals. For example, an informal review of a new teaching practice may be conducted once the practice is implemented to ensure students are benefitting.

A formal review provides credible feedback for sponsors, supporters and critics, parents, and the community. Transparency is the key.

Essential Element 1

Questions & Actions

🔷 **In your plan, outline your process and those responsible for quality assurance.**

1. Establish a review group to oversee progress and resolve problems.

Select someone to be responsible for the ongoing evaluation process. Although all key team members are responsible for ongoing evaluation, this person will ensure the process is in place and that appropriate information and feedback is not only gathered but that there is a process in place for analyzing the feedback and making any necessary changes and adjustments in the initiative implementation. There are always opportunities and a need for adjustments.

2. You may want to review:

- **Portugal Transforms Primary Education with 1:1 Technology Integration**,[B52] which describes how Portugal developed its 1:1 initiative around both clearly defined learning and economic goals and the monitoring process it put in place.
- **Quality Assurance: Monitoring and Evaluation to Inform Practice and Leadership**.[B53]

Essential Element 1

3. **Decide on a formal review process that is clear to all the stakeholders.**

   It should be based on the goals and outcomes outlined at the start of the initiative and any sub-goals developed from these. It should include a formal process for developing actions based on the review group's recommendations.

4. **Look at each component of your initiative's plan. Develop answers to the questions in the tool entitled Designing Evaluation Indicators in Part III - Tools and Resources.**

   Develop a series of questions like these around each of these areas that will be part of your evaluation process. Outline your evaluation process in your initiative plan.

5. **Set up a schedule for reviews of the evaluation outcomes and for implementing adjustments. Add your schedule to your initiative plan**

6. **Review and evaluation are an essential and ongoing processes. Continue having regular reviews as your initiative matures.**

PLANNING OUTCOMES

As a result of your research, readings, and reflections…..

- Outline your mission statement and goals here.
- Ensure school, district, and any national goals are aligned.
- Include milestones.
- Outline your evaluation process and how feedback will be used.

Essential Element 1

Specific 1:1 Goals

The national and/or district 1:1 goals, outcomes, milestones and the assessment measures that will be used to measure the levels of goal achievement are:

Year	Goal	Outcome	Milestone	Goal assessment measure
1				
2				
3				
4				
5				

FOR SCHOOLS: Our school 1:1 goals (aligned with national and/or district goals), outcomes and milestones and the assessment measures that will be used to measure the levels of goal achievement are:

Year	Goal	Outcome	Milestone	Goal assessment measure
1				
2				
3				
4				
5				

National or District Policies to Consider

Evaluation Team Leader

Evaluation Process and Plan

Additional Resources Used

Essential Element 1

Essential Element 2 - What's Now Possible: Teaching & Learning for Contemporary Learners

PRINCIPLE: Learning in a digitally-rich world requires the development of new pedagogies to best leverage emerging technologies' potential to amplify opportunities for today's learners.

> We need school to produce something different, and the only way for that to happen is for us to ask new questions and make new demands on every element of the educational system we've built. Whenever teachers, administrators, or board members respond with an answer that refers to a world before the rules changed, they must stop and start their answer again.
>
> No, we do not need you to create compliance.
> No, we do not need you to cause memorization.
> And no, we do not need you to teach students to embrace the status quo.
>
> Anything a school does to advance those three agenda items is not just a waste of money, but actually works against what we do need. The real shortage we face is dreams, and the wherewithal and the will to make them come true.
>
> No tweaks. A revolution.
>
> — **Seth Godin,** *Stop Stealing Dreams (What Is School For?)* [C1]

What does ubiquitous access to technology make possible for contemporary learning and teaching?

Essential Element 2 focuses on rethinking and redesigning all aspects of teaching practice and the learning experience, based on the information garnered and ideas developed in Element 1. With a well-defined vision and actionable goals, it is now possible to boldly analyze and shape all aspects of learning and teaching practice in order to achieve these goals and vision.

Learning and teaching practice must guide all thinking throughout the planning, preparation, and implementation of a 1:1 initiative. Educational leaders now have a unique, unprecedented opportunity to forge new directions for their schools and for the experiences of their young learners.

In **Essential Element 2**, district and school leaders, including principals, work with the Director of Curriculum and Instruction, teacher leaders, and technology integration specialists to rethink all aspects of the learning process and environment and the impact this has on the role of teachers.

Project Team Members

- District and/or school leaders, including principals. • Director of Curriculum and Instruction.
- Academic Director. • Teacher leaders. • Technology integration specialist.

Essential Element 2

Explore Contemporary Learning

> ..the true power of the computer is that it is capable of manipulating not just the expression of ideas but also the ideas themselves.
>
> The computer is not just an advanced calculator or camera or paintbrush; rather, it is a device that accelerates and extends our processes of thought.
>
> It is an imagination machine, which starts with the ideas we put into it and takes them farther than we ever could have taken them on our own.
>
> — **Danny Hillis,** *The Pattern on the Stone* [C2]

Time and again, schools report that the deployment of 1:1 devices in the absence of any other changes has had limited or no impact on improving learning outcomes. Why would anyone find this surprising?

Too often when we introduce technology, we are looking back, at our existing practice, and overlaying technology on top of it; or at best, we are 'integrating' it into our existing practice.

It's time to look forward.

Essential Element 2

Questions & Actions

> How will ubiquitous technology change how students learn? For example, how does the availability of diverse, high quality and easily accessible content or easy online publishing change the way our students learn? How do these changes impact how you assess learning? What are the prerequisites for a modern learning environment?

It is now possible to rethink all the 'sacred cows' of education, including:

- Assessments
- Separate academic departments
- Grade levels organized by age groups
- Timed tests
- Use of time: schedule and calendar
- The boundary between academic and co-curricular and/or informal learning
- Multiple choice true/false tests and quizzes
- Individual achievement assessment vs. group assessment
- Equating learning space to a classroom

Essential Element 2

> *A truly responsive and affirmative education is not just about how many subjects we offer (like goods on a shop shelf); it is about learning to know, respect and trust the talents and energies kids bring into class.*
>
> *Has our obsession with neat boxes and crude rankings now quite obscured any lively sense of community, mutuality and shared enterprise?*
>
> *We need to consider the inescapable long-term result of schooling all young people all day long in peer-group cohorts. Is not the primacy of peer-group pressures being perfectly engineered by our streaming, grading and sorting? Few cultures have ever made such absolute use of age-banding their young.*
>
> — **Michael Norman, former Principal of Woodleigh School, Australia,** *A Need to Cater For All Talents* [C61]

1. Further to the above, discuss the following "exercise of the educational imagination":

> *As an exercise of the educational imagination to strengthen your visionary powers, think about a world in which there is:*
> *No such thing as fourth grade, because age segregation has gone the way of other arbitrary divisions of people.*
> *No such thing as a classroom, because learning happens in a variety of settings.*
> *And no such thing as curriculum, because the idea that everyone should have the same knowledge has come to be seen as totalitarian.*
>
> *Your assignment in this exercise is to figure out whether such a world could work. Be clear. I am not proposing this as an education reform. .. I am doing something different in kind from proposing or predicting: I am suggesting that seriously developing and seriously confronting alternative scenarios be recognized as a valuable kind of work—work needed to facilitate the emergence of the future.*
>
> — **Seymour Papert,** *Let's Tie the Digital Knot* [C3]

Essential Element 2

ideasLAB, an Australian education think tank, identifies three dimensions of contemporary learners; they are connected (social learners), curious (inquirers), and self-directed.

The ultimate aim of education is to enable individuals to become the architects of their own education and through that process to continually reinvent themselves.

Elliot W. Eisner, *The Arts and the Creation of Mind* [C13]

1. Review the ideasLAB white paper **Understanding Virtual Pedagogies for Contemporary Teaching and Learning.** [C4]

2. The **12 Principles of Contemporary Learning** are based on the three dimensions mentioned above.

   Consider how each of these principles provide new opportunities for learning and teaching, as well as new challenges. Begin to outline new ways learning can be organized to reflect these dimensions.

Essential Element 2

12 Principles of Contemporary Learning

*Through the lens of the **social learner**...*

1. Modern learners have the ability to access high quality content whenever and in whatever format they need it...which enables them to draw upon a diverse range of external resources.
2. Modern learners have the ability to publish using a variety of media for low or no cost ... which enables them to share their ideas and get feedback from others.
3. Modern learners have the ability to form networks ...which enables them to contrast ideas and experiences with other learners.
4. Modern learners have the ability to form highly interconnected groups around an object of interest...which enables them to engage in shared meaning making and work collaboratively.

*Through the lens of the **inquiry-based learner**....*

5. Modern learners have the ability to save and retrieve information in a variety of formats ... which enables them to extend their capacity to manage and manipulate information.
6. Modern learners have the ability to work in teams to participate in open and distributed projects...which enables learning teams to manage more complex projects.
7. Modern Learners have the ability to reuse and build upon the work of others ...which enables them to move beyond individual and isolated projects.
8. Modern Learners have the ability to quickly obtain feedback from multiple sources ...which enables them to continuously improve their current work.

*Through the lens of the **self-directed learner**...*

9. Modern learners have the ability to generate large amounts of data about their technology-based activities...which enables them to use self-generated data to assess and make decisions on future actions.
10. Modern Learners have the ability to view the learning artifacts of others ...which enables them to learn from what other learners are doing or have done.
11. Modern learners have the ability to amplify important and noteworthy content ...which enables them to easily identify and participate in new and noteworthy ideas.
12. Modern learners have the ability to operate in the same spaces as experts and professionals...which enables them to make better decisions about their own learning.

...the remarkable feature of the evidence is that the biggest effects on student learning occur when teachers become learners of their own teaching, and when students become their own teachers. When students become their own teachers they exhibit the self-regulatory attributes that seem most desirable for learners (self-monitoring, self-evaluation, self-assessment, self-teaching).

John Hattie, *Visible learning: A synthesis of over 800 meta-analyses relating to achievement* [C60]

How will each of these have an impact on how learning is taking place and the roles schools will have in this new environment of learning? What questions do they raise?

1. To get started, consider and answer the following questions:
   - How does the availability of diverse, high quality and easily accessible content change the way our students learn?
   - Is it really the teacher's role to quality assure online content?
   - How does easy online publishing change the way our students learn?
   - How do schools respond to students as autonomous, self-directed learners?

2. What other questions do these **12 Principles** raise in terms of creating the best conditions for contemporary teaching and learning?

> It's the change underlying these tools that I'm trying to emphasize.
>
> Forget blogs...think open dialogue.
>
> Forget wikis...think collaboration.
>
> Forget podcasts...think democracy of voice.
>
> Forget RSS/aggregation...think personal networks.
>
> Forget any of the tools...and think instead of the fundamental restructuring of how knowledge is created, disseminated, shared, and validated.
>
> — **George Siemens,** *Connectivism*[C5]

3. Use the following questions along with the videos to provoke discussions across your school community about how our young people are learning today in a technology-rich world:

- What does unlimited capacity make possible for your students?
  Watch **Is Your Phone Part of Your Mind?** (David Chalmers, TEDxSydney) [C6]
- How would you best define 'modern' literacies?
  Watch **New Media Literacies** (Project New Media Literacies) [C7]
- How should we be teaching mathematics in a technology-rich school?
  Watch **Stop Teaching Calculating, Start Teaching Math** (Conrad Wolfram, TEDGlobal 2010) [C8]
- What are the benefits of moving from individual authorship to co-authorship? How might we rethink assessment in this context?
  Watch **Embrace the Remix** (Kirby Ferguson, TED talk) [C9]

> *Substantially all ideas are second-hand, consciously and unconsciously drawn from a million outside sources.*
>
> — **Mark Twain**

4. As an extension of the **12 Principles** above, here are just a few of the applications to which students now have unlimited access. Explore the following links and discuss their role in learning today and in the school learning environment.

- **Fanfiction.net**
- **Delicious.com**
- **Wattpad.com**
- **Scratch.mit.edu**
- **Stackexchange.com**
  - math.stackexchange.com
  - physics.stackexchange.com
- **KoduGameLab.com**

Essential Element 2

5. Watch and discuss the video **Science Leadership Academy**.^{C10} How do the student projects reflect the **12 Principles**? Consider the following:

Don't assume you know every outcome. You don't. We don't. And that's okay. That's a feature, not a bug.

Chris Lehmann, Science Leadership Academy

For many policy makers and administrators, buying thousands of iPads for a school may seem like a progressive move, but a generation that has already grown up with these devices will expect them to be more than just a replacement for hardcover books.

Stuck Between a Book and a Hard Place^{C16}

6. The following blogs provide insights and ideas around how learning happens:
    - **Games Based Learning**^{C11}
    - **Project Based Learning**^{C12}

Select several posts from each to discuss as a team and explore how the ideas expressed might be applied at your school or in your district.

7. Consider this statement — "Every classroom should be considered a 'maker-space'." What does this imply? How does this impact your thinking of how learning takes place?

Essential Element 2

> *Making is predicated on the desire that we all have to exert agency over our lives, to solve our own problems. It recognizes that knowledge is a consequence of experience, and it seeks to democratize access to a vast range of experience and expertise so that each child can engage in authentic problem solving.*
>
> **Gary Stager,** *What's the Maker Movement and Why Should I Care?* C13

8. In **Essential Element 1,** you explored expectations for what 1:1 makes possible – from both your perspective and that of your students. Based on the **12 Principles**, carefully consider what the impact will be on your expectations of the capabilities of young people.

Essential Element 2

Critical Conversation: Anytime, Anywhere Learning

When we think of learning, we tend to focus on formal learning, that which takes place at school and is led by one or more instructors. This formal learning historically centers around subjects and specific bits of information/knowledge that could not be readily obtained outside the formal educational setting, delivered to students from knowledgeable adults with unquestionable expertise. What is of importance to be learned is what these adults intend for the students to learn and assessments are based on whether or not they have succeeded.

On the other hand, we all recognize that learning doesn't just happen at school. We are all learning, all the time. What happens outside of school is life, hobbies, or, for many older students, work. In these activities, young people may have the opportunity to practice a variety of skills, from sports, to music, to more active social and interpersonal skills. Often these are in supportive, encouraging environments because learners in these contexts are usually also consumers, with the ability to choose to go or not to the activity.

So the role of the mentor/coach or learning designer is to structure learning so that the learner is both successful and engaged in the process. Even students with after school jobs are actively gaining both explicit and tacit knowledge, within a role that encourages them to assume responsibility that impacts not just them but their colleagues.

Too often, though, we don't recognize or acknowledge the process as a form of learning, nor do we place much value on most of that learning, which is often of a practical or more informal vein. It is seen to have less value within the commonly accepted idea of what it means to be educated (although when practical knowledge is what is needed, having only 'school-smarts' is seen as a disadvantage).

In this age of ubiquitous technology, where, as Don Tapscott states in **Wikinomics**,[C63] "As more and more universities and other schools put their courses and course content online, there will be no dearth of content and no requirement that the school be the source of content," these distinctions between what one learns at school and what one learns outside of school begin to fade. We no longer need to divide the day, the week, the year, between 'learning' and 'living,' between learning that is considered more valuable because it's assessed with a test vs. learning that can be demonstrated and connects to day-to-day life, the community, and the world. Rather we can truly develop the idea of lifelong, anytime, anywhere learning.

If we were to consider all our waking hours as learning time, how would we make best use of the time we meet together at a school with pedagogical experts? And what would be the role of these experts?

> Today's schools, as they are most commonly organized, were designed for an earlier time and place, with different goals in mind, and when less was known about what the optimal conditions for learning are. But their design is not immutable.

1. Discuss and answer the following three questions:
   - **What do we keep?** (What practices are essential to keep?)
   - **What do we add?** (What new practices will help us achieve our vision and goals?)
   - **What do we throw out?** (What practices must we stop doing?)

   Begin to rethink all aspects of learning and the learning environment using the new opportunities 1:1 creates. Involve your Director of Curriculum or Academics (or whoever is in a similar leadership position) in these discussions and you may want him or her to initiate the necessary research and design work.

   (Although all these questions are challenging, often the most difficult one to act on is the third. It is not enough to add new practices on top of old practices that do not provide the support needed for a modern learning environment.)

2. A fourth question must also be considered:
   - **Why?** How do we know what we should keep or throw out?

   The criteria need to be focused on whether or not a practice has a positive impact on learning (which implies you also need to come to a shared understanding of what constitutes a 'positive impact').

> **What is personal learning? How is it the same or different from personalized and/or differentiated learning?**

Essential Element 2

 *Personal learning entails working with each child to create projects of intellectual discovery that reflect his or her unique needs and interests. It requires the presence of a caring teacher who knows each child well.*

Personalized learning entails adjusting the difficulty level of prefabricated skills-based exercises based on students' test scores. It requires the purchase of software from one of those companies that can afford full-page ads in Education Week.

Alfie Kohn, *Four Reasons to Worry About "Personalized Learning"* [C17]

1. Do you agree with how Alfie Kohn defines personal and personalized learning in **Four Reasons to Worry About "Personalized Learning."** (Consider reading the complete blog from which this quote is taken.)

   What are the obstacles to 'personal' learning and how can they be tackled?

2. Watch **21st Century Learning in Action**.[C18]

   Although mentioning a variety of ways 1:1 impacts learning, this video particularly emphasizes the dimension of self-directed learning, including meaningful student voice.

   Discuss with your pedagogical leaders how this shifts the learning environment.

3. Watch the video **Bow Drill Set**.[C19]

   The young boy in this video exemplifies self-directed, passion-based learning. His use of YouTube highlights not just the use of technology, but the ability to connect to a community of knowledgeable others in a very careful, thoughtful way.

   Discuss how to define personal/ized learning based on this video. How does this compare to your previously held ideas about personal learning? How can these ideas be applied to your schools?

Essential Element 2

4. Read and discuss **Personalized Learning for Global Citizens**.[C20]

As a team, write your definition and beliefs around personal learning.

> How will you give students a more meaningful voice in their learning?

---

from **Students as Agents of Change**[C21]
Sylvia Martinez, former Executive Director, GenYes

Implementing a laptop initiative often involves months, sometimes even years, of planning and is usually steered by stakeholder groups consisting mostly of teachers, administrators and parents. **But students are 92% of any school population. This is an often overlooked yet crucial stakeholder group.** Including students on planning committees creates a larger sense of ownership of the program once it becomes a reality. Ongoing student input and support of laptops provides valuable, knowledgeable resources, additional team members, and student point of view.

There are two basic ways students can participate and contribute to the planning and implementation of a laptop program:

Committees: Students can participate in technology planning committees, school site councils, technology security committees, or peer review committees. Adults often claim that including students in planning is risky, citing privacy concerns, lack of maturity, or difficult logistics. However, adults often forget that accommodations are made for them when they are included in such planning committees: adults may or may not know anything about technology; they have schedules to work around, and may not have been in an actual classroom for years if not decades.

Having students participate in committee work is not only a wonderful learning opportunity for the student but creates a direct path for student feedback and point of view that is extremely beneficial to teachers and other adults. Adult guidance is key to making this student participation successful, since otherwise students may find the meetings long and tedious. A great way to prepare students for such meetings is through role-play. Meet with students regularly prior to and after meetings to discuss progress and get their feedback.

Day-to-Day Activities related to Laptop Support: Include students in various roles supporting laptop use. This can include basic technology support, instructional support, or helping new users learn about their laptops. Student help can make the logistics of implementing your program run more smoothly and may also ease new users' anxieties. Assemble a student tech team before you begin your program in order to train them on the hardware and software, as well as to familiarize them with new policies. Once the laptop initiative becomes a reality, you will have an enthusiastic, trustworthy student team raring to go.

Essential Element 2

1. Include students in the discussions and planning as you rethink their learning environment. Make them part of the process.

2. Explore **TakingITGlobal**[C22] to better understand the projects they support and how they provide young people with authentic opportunities to develop the skills they need to make decisions about and impact the world in which they live.

3. A few schools have created a formal mechanism that allows and supports students as partners in their own learning. Read and discuss the UK report **Learning About Learning**.[C23]

Essential Element 2

Critical Conversation: More Than Just Facts

In its zealousness, the institution of School has gone overboard in assembling information - if a few facts are good, a thousand must be better. And in this over-enthusiastic effort to provide more and more content, schools have forgotten their goal while drowning their students in data, detail, and Jeopardy answers. Teachers have had to swim faster and faster to keep afloat with the ever rising river of content, especially as the amount of information is growing exponentially, leaving them little time to spend reflecting on the bigger picture.

If, on the other hand, we assume there are numerous sources of facts and content available digitally, as is assumed by people working in all other industries (and by industries, we're being all-inclusive, so not just revenue-generating entities, but also non-profits, the arts, social services, etc), then not only can School do what we think it should be doing, but the role of the educator would be radically different.

This is not just flipped learning, which frequently still has too much of an emphasis on content during school time as well as in terms of assessment. And it isn't about spending class time doing individual projects or even co-produced projects that don't entail real collaboration. This is completely rethinking what happens after the facts are gathered - and spending all of school time doing that, doing the learning. Once again, technology plays a key role, providing new ways to explore big ideas extrapolated from the information each learner has assembled, connecting and exchanging views on different interpretations of 'the facts' from different perspectives.

It may be best at this time to stop talking about School as one continuous, identical process and divide it into three phases - early years, middle years, last 2-3 years - similar to what exists, but with the majority of students in the middle phase. Note, this is not elementary, middle, and high school, since the two ends consist of only small periods of time, with the first focused on literacy and retaining children's natural enthusiasm for learning and the final couple of years focused on aligning with the more conservative structure of universities.

Once students are competent and happy readers, they are ready to enter into environments reflecting a new structure built around providing students with unlimited opportunities to learn and within which educators must assume new roles.

How can you redefine curriculum to be a strategy for deeper learning? How will ubiquitous technology change what students learn? By shifting from teacher to student-directed learning, what first steps can you take to develop a contemporary curriculum strategy?

1. Review the information at **Innovative Teaching and Learning Research**.C24

2. Create a shared understanding of what 21st century skills are. Be clear about your definitions.

Essential Element 2

So, what really are 'the basics' of mathematics, those core ideas and skills all people need?

Is calculation really the most important skill?

I realize that defining 'the basics' is not a simple task, but shouldn't we feel the necessity of doing this in light of the tools that are currently available and that have re-shaped so many aspects of our lives? Skills that were deemed absolutely essential when the tools we had were different may no longer be the skills that are necessary today.

— Rethinking Math Basics in a Digital Age [C25]

3. Read and talk about the ideas in the report **Connected Learning**.[C26]

Consider this statement from the report:

Connected learning … is not simply a 'technique' for improving individual educational outcomes, but rather seeks to build communities and collective capacities for learning and opportunity… Without this focus on equity and collective outcomes, any educational approach or technical capacity risks becoming yet another way to reinforce the advantage that privileged individuals already have.

How would you define connected learning? Why is it important for learning today? How can you apply the ideas presented in this report to your school?

Essential Element 2

from **How 1:1 Enlivens Math Classrooms** [C27]
by Brandon Dorman, Teacher,
Computech Middle School, USA

A watershed moment about the power of 1:1 environments in math came towards the end of last year when my team used a one minute video from Dan Meyer's 101qs.com about a domino spiral. Students only saw the outer ring fall, but had to use all of that information to find out how long it would take for the entire pattern to fall. We let them use their own devices to measure how long it took, to time, to chart, to collaborate, and many students utilized all of the tools we had previously learned with! We were thrilled to see students using Geogebra to make a model, Google calc to input data into a spreadsheet to predict future times with formulas (circumference, rate formulas...), and most importantly, students playing the video over and over to see new patterns and predictions.

Since students had their own devices to view the data and video, they were able to all take a look at different ways to investigate the problem. In all, students found seven distinct methods to approximate the time it took for the dominoes to fall that were within .5 seconds of the actual time it took.

The class time was not without its challenges. Many students wanted to give up or have me give them the answer. To remedy this I set up a class Padlet page (a digital bulletin board - students like the free-form nature of the board as opposed to a stale google Doc) where students asked each other leading questions that had helped them find the answer.

Without a 1:1 environment in use, students would have passively watched the video and probably come up with one or two methods of presenting their findings. Having laptops and tablets in front of them with the video enabled the students to take the problem and make it their own, with their own paths to a solution. In addition, the classroom was abuzz as students struggled to find different solutions, and it was amazing to hear how one possible solution would start with one group and begin to ripple across the rest of the room and even between different classes. One to One devices in a math classroom enable shy students to not be afraid to fail, and more importantly allows the teacher to give direct feedback to students to re-engage the task from another angle.

PLANNING OUTCOMES

As a result of your research, readings and reflections...

- Discuss and begin planning for shifts around when, where, what, and how students learn.
- Explore new forms of assessment and the impact this has on how students learn.

Embrace New Roles for 21st Century Educators

> I've changed the way I see myself as not a teacher, but an architect, in charge of creating a space for the learning, setting the goals and then seeing how we move through that. And that can be different, not every girl has to move through my curriculum in exactly the same way. In this way I can foster individual passion...
>
> ...What crystallized in my mind, the lightbulb moment, was that these kids have access to the same information that professionals do. So they should be doing the same things that scholars are doing with that same information.
>
> I truly think what I want is for them to walk out of my classroom and participate in society, have an opinion about what's going on around them, and have an opinion about the kind of world they want to live in.
>
> — **Carolyn Thompson, Teacher, McGehee School, USA** [C28]

Modern learning environments redefine the role of the educator. This can be overwhelming, but it also offers new opportunities for teachers to explore what is now a much wider range of possibilities for teacher-learner interaction. For one, the teacher no longer has to assume sole responsibility as the deliverer of content and information, a time consuming task that reduces the time an educator has available to explore and implement methodologies for deeper learning.

Essential Element 2

Questions & Actions

What are the emerging roles of the 21st century educator?

1. **Re-examine the role of teachers to determine how best to support contemporary, anytime anywhere learners.**

    Educators are now mentors, coaches, advisors, learning strategists, talent scouts, networkers, curators, and researchers exploring their own teaching practices. Discuss as a team what these new roles entail.

Essential Element 2

Critical Conversation: Teacher as Talent Scout

We all have met young people, whether now or in the past, who have an unusual passion or talent for a sport, or music, or some subject. If the school does not support them, we think they have missed an opportunity to know that learner and make his/her learning meaningful. If the talent is sports, there is frequently support at school, but in other areas, recognition and support of talent can be spotty, uneven, or non-existent.

Too often teachers, especially in secondary schools, where large numbers of students are processed, do not get to know their students enough to know about these passionate interests in a field. Usually, communication is mainly unidirectional and focused on the topic at hand. There is little time to do otherwise, and no precedent to think it should be otherwise.

Although not all young people may identify their interests as passions, they all will admit to some interests. It is, as Sir Ken Robinson says, understanding what subjects, topics, or ideas spark that interest in a student and not ignoring that spark to instead focus on a curriculum that may not include that area that sparked a response. And what if, while discussing ideas related to the curriculum, nothing sparks interest?

Then it is time to realize that there are a number of ways to explore the big ideas that run through the major disciplines, and those that may have been more narrowly defined as the 'curriculum' may not be the appropriate ones. Take science. It need not be only chemistry, physics, and biology (with occasionally some ecology and perhaps some robotics thrown in) - but it could be offered for some via a different approach that not only triggers interest, but integrates a variety of disciplines, is more linked to a student's life, may be more impactful, and that requires thoughtful, rigorous work to understand.

For example, some students may be interested in the exploration of genetically modified foods, others in archaeology and carbon-dating and its difficulties, others in designing racing cars or sports equipment with a winning edge. Each of these topics has rich links to so many different disciplines and none are trivial or simple.

So, in being a talent scout, the teacher, now more of a learning strategist, must be able to connect with his or her students and recognize that spark of interest. It is not a difficult task, but it does take knowing how to ask the right questions, observe reactions, and draw out the reasons behind or connection to the topic. This ongoing dialog can help shape the exploration and provide authentic connections to a variety of disciplines. Often, students, given the opportunity and encouragement to talk about their interests, will do so gladly.

For students not used to being asked where their interests lie, the process to elicit the information should be not only risk-free (no ridicule, no right-or-wrong answers), but may need to begin broadly with general questions that help them think about what their interests are. The answers to the initial questions help determine the follow up questions, which should help the student more precisely define his or her interests, without narrowing them too much.

Essential Element 2

Students also need to be exposed to a broad range of new ideas and topics, and school is the ideal place to do this. Focusing on a narrow curriculum may work for a narrow group of students, but it will not necessarily appeal to a broad range of students.

Schools need to find opportunities to introduce topics not normally considered part of the 'basic package' curriculum developed two centuries ago. They need to have the opportunity to hear and learn from a broad spectrum of people, be exposed to a variety of ideas, and then work with an educator who understands how to recognize when and what triggers some intellectual excitement in each student, and how that leads to deep, personal learning.

What is personal learning? It is not just data and tracking and redirecting through programmed learning that is just a digital version of traditional textbooks and worksheets. It is knowing each student and engaging in conversations with them as well as paying attention to the data, establishing relationships between the adults and the young people in the school, including providing digital tools that increase communication, enhance collaborative work, create networks, and allow each learner to follow a path that connects to his or her interests, level of understanding, and to the bigger questions that should be explored in a 21st century curriculum.

Essential Element 2

2. Discuss the following two research reports:
   - **Transforming Learning Environments for Anytime, Anywhere Learning for All**[C29]
   - **Curriculum, Content, and Assessment for the Real World**[C30]

3. As technology and 1:1 change our thinking around curriculum, content, and assessment, how do these changes impact the role of the teacher?

> …. in 1990, how did it affect an 11-year-old girl to see her forty-plus year-old female teacher eagerly teaching herself new technology skills, innovating in her craft, and revelling in ambiguity as an opportunity to learn? Might that student's attitudes towards learning then, and even today as an adult, be partially the result of this hidden curriculum? And if so, what is the message sent by, and what is the cost of teachers who act antithetically? Today, I ask teachers to see their own attitudes towards change, new learning, and coping skills as part of the curriculum.
>
> **Adam Smith,** *Reflections on 20 years of 1-to-1 at Methodist Ladies' College*[C31]

Essential Element 2

4. Look at what schools around the world are doing as they embrace new roles for educators. Analyze how each of these schools has defined the role of their teachers. What roles will teachers in your school have? How might you incorporate some of the ideas from these videos into your learning environment?

- Listen to school leader **Dan Buckley** [C32] from Saltash.net Community School in the UK talk about how he enables his educators to place the student at the center with Personalized Learning, then read **Saltash.net's Case Study**. [C33]
- Watch how **Colegio Fontan School** [C34] in Colombia eliminated classrooms and succeeded in competency based learning, then read **Colegio Fontan's Case Study**. [C35]
- Hear school leader **Larry Rosenstock** [C36] from High Tech High in the US talk about project-based learning, then read **High Tech High's Case Study**. [C37]

How do these changes impact classroom culture?

5. Watch **New Roles for 21st Century Educators**. [C38]

In this video, educators from around the world describe the new roles they have as educators in the 21st century when developing the type of learning environments 1:1 makes possible.

Essential Element 2

> Sometimes a student will correct you as you speak based on a search they just did. I think those are great teaching moments. It undermines formal intellectual authority and shows that the search for knowledge is equally open to everyone.
>
> **Rick Salutin,** *Don't Ban Those Laptops in Class*[C39]

In what ways will changing the role of teachers also change how they work (for example, teachers may no longer work in isolation, but rather in teams, etc)? In what ways will teachers use ICT?

> The problem is that school is incredibly boring for students, and as it turns out, it's pretty boring for teachers as well, because neither group is learning in the traditional school model.
>
> **Michael Fullan**[C58]

Essential Element 2

Critical Conversation: Learning Teams

Currently in schools, all teachers are expected to be experts - not merely acquainted with, but experts - in the whole range of skills needed to help students achieve their learning potential. And each teacher is expected to have this same range of expertise. It is an unusual expectation. This is like having a company with 30 employees, and instead of having people with knowledge in the areas of finance, marketing, administration, and development, each employee would be hired to do the exact same thing: everything. Furthermore, any company in the same industry would be peopled by the same type of employees. Absurd.

Since realistically a teacher cannot be an expert in all things, teaching in many cases and places has been reduced to mean "provider of content" or "the expert in the classroom." Teachers present content, administrators administrate - matrix management at its simplest.

Who would I want in my Learning Team?

Just like a team in a hospital, students should have a 'learning team' led by one educator, but with knowledgeable contributions from a number of other 'specialists' - clearly content specialists, but others who may have a specialty in a particular form of expression (kinesthetic, visual, musical, etc.) or a particular passion that is not necessarily part of the 'typical' curriculum plan (for example, archaeology, architecture, radio production, etc.). And, of course, the learner is part of the learning team, and not an object upon which they operate (in contrast to what happens in a hospital). The role of each person on the team depends on his or her areas of expertise, so each educator would not be expected to be all things to all students, but rather the ensemble of educators as a whole would have the necessary knowledge and skills that the school has deemed necessary.

School leadership needs to create an environment that allows these teams to form, function, collaborate, and continue to analyse, reflect, and learn from each other in an ongoing, iterative, growth process.

Although each of these specialists contributes to the learning, it is the task of the lead educator to take the lead as learning strategy advisor and help create a cohesive, relevant learning program for and with the student that meets the student's learning goals.

In order for the lead educator and the team to be most effective, it is essential that they know their students - their interests, their strengths and weaknesses, their goals, and not just their grades and test scores. This knowledge would come from educators' observations, conversations with each student, plus tools designed not just to highlight academic strengths and weaknesses, but to help students explore and articulate their personal interests.

Essential Element 2

1. Read and discuss **Exploring the Future Education Workforce: New Roles for an Expanding Learning Ecosystem**.[C40]

2. What other new roles do you envision for teachers? How will these change how they work?

> **Ideas can come from many different places and industries.**

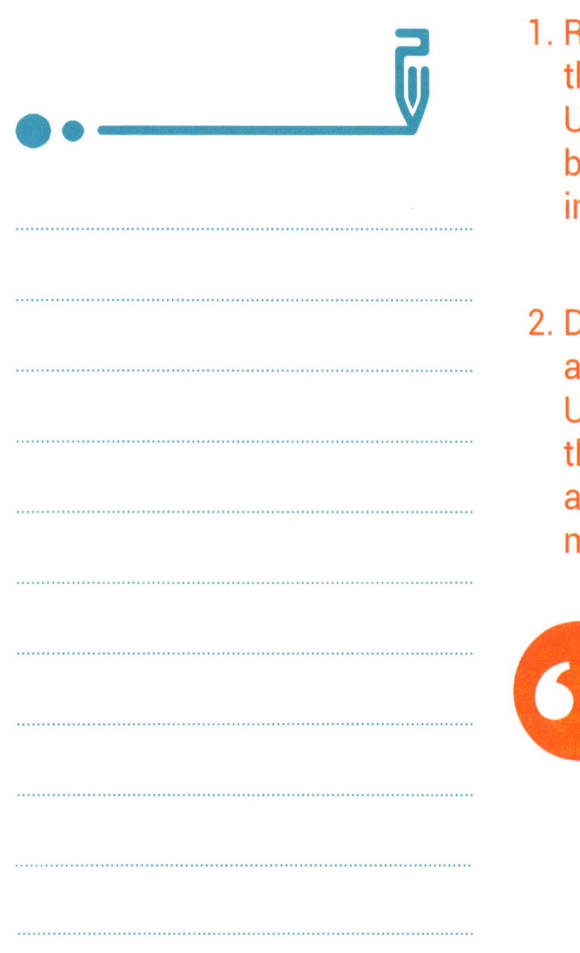

1. Read **The Battle Is For The Customer Interface**[C41] that describes how Uber and companies like Uber have very successfully targeted the layer between customer and supplier – the all-important interface layer.

2. Discuss how these ideas could link to education and school. What can educators learn from Uber? Is there a new role for educators in the layer between the abundance of content available and the student? What could this mean?

> *It is not sufficient to equate 'being in charge of one's own learning' with 'student agency'. Kids can not be in charge of their own learning if they do not understand the intricacies of learning and, indeed how they learn in different circumstances.*
>
> — **Peter Skillen,** *So You Want Kids to Code! Why?*[C42]

Essential Element 2

What are the emerging/essential roles for 21st century principals? In rethinking the roles of both learners and teachers, it should come as no surprise that the principal's role has also shifted. Through the last century, this school leader's role has become more and more administrative with less and less time to be the 'principal teacher' and pedagogical leader. Will there now be new opportunities for school leaders beyond that of administrator?

> *School has changed from being a daily chore to space for thinking, expression and community. I am witnessing teachers experiencing a new role – no longer the managers of behavior, but now the creative directors of curriculum delivery.*
>
> **Stephen Harris, Principal, Northern Beaches Christian School, Australia,** *Daily Edventures* [C43]

1. In considering how learning and teaching are changing, what different roles may also emerge for school leaders? How might this shift their priorities as school leaders?

   For example, the hierarchical order of the institution is giving way to one of shared leadership. What might be the outcomes or consequences of such a change on the role of the principal?

2. How can school leaders best model new opportunities for learning, connecting, collaborating, and creating that ubiquitous access to technology enables?

> *Thinking big is not tinkering around the edges, but creating a different mold and facilitating transition to the new mold. There are many different designs to support the creation of experimental spaces, new roles, etc. However, these require each of us thinking and acting like "leaders", and not managers.*
>
> **Shabbi Luthra, Director of Research and Development, American School of Bombay, India,** *Leaders or Managers, What Do Schools Need?* [C63]

Essential Element 2

from **School Leaders and 1-to-1: Model Behaviors** [C44]

by Dr. Paul Fochtman, former Superintendent, American School of Bombay, India

A note for school leaders: Technology integration has to start with us. We need to model behaviors that support the integration of technology.

Some ideas include:

- Communicating the critical importance of integrating technology to staff and board;
- Remaining a learner--staying current with technology changes in the real world and being abreast of emerging technologies and their place in education. Exploring the use of emergent technologies and encouraging healthy risk taking;
- Participating in training with the teachers or in visible training that enriches your role;
- Modeling the use of electronic communications;
- Balancing polarizing views of technology by facilitating pace, sharing quantitative and qualitative success, and moving educators past the "creative tension" stage. (Senge)

It is no longer necessary to apologize for the importance and necessity of technology in our lives, but rather this is our opportunity and responsibility to embrace technology. Educators will need to hear from school leaders that technology, when used effectively, will result in data which clearly impacts student learning and achievement.

The process of integrating technology into the classroom in ways that are transformative is a challenge. The end result is an enriched teaching and learning environment in an era that measures traditional skills but requires 21st century skills. As leaders, we need to recognize the landscape is changing and we'll need to address both.

Essential Element 2

PLANNING OUTCOMES

As a result of your research, readings and reflections…..
- Describe what learning and a modern learning environment will look like in your school in your initiative plan.
- Outline what you envision as the new roles for teachers and what this means in terms of best practices and learning.
- Outline what you envision as the new roles for principals.
- List your selected virtual learning space tool(s) and briefly describe why you are recommending these.
- Create a list of essential and recommended changes to the physical learning environment here.

Our beliefs around learning and teaching are:

Addressing the dimensions of learning…

…….Our connected learners will…

…….Our collaborative learners will…

…….Our inquiring/curious learners will…

Personalized learning means…

In our school, personalized learning will look like this…

From these beliefs, we see the following shifts happening…

The role of the teachers in our school(s) will be…

In order for curriculum to be a strategy for deeper learning, we will…

The shift from teacher to student-centered learning means we will…

Assessment will look like…

……because………

Constraints and challenges to overcome include…

The role of the school leader(s) in our school(s) will be…

Additional Resources Used

Essential Element 2

Design a Modern Learning and Teaching Environment

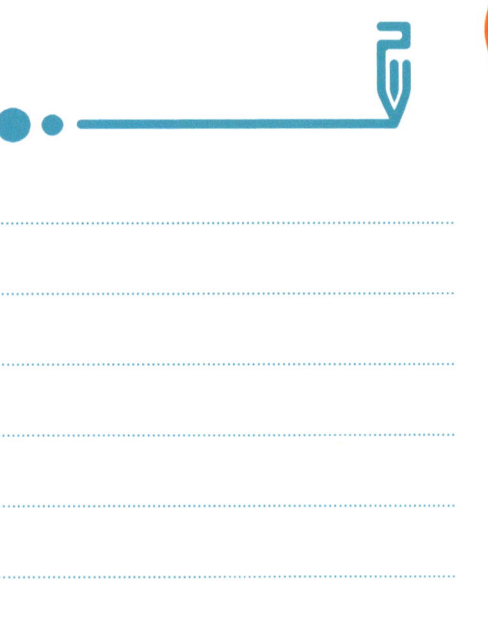

Image Credit: © Silverton Primary School, Australia

> *If schools are about learning, shouldn't school architecture also be about learning? And if school architecture is about learning, shouldn't it foster certain pre-conditions for learning such as thermal, acoustic and physical comfort, good lighting and acoustics? Shouldn't school architecture support various learning modalities, allow its users to constantly tailor and modify their space to meet changing needs? Shouldn't it, most of all, be about what children need for their special developmental needs at various stages in their lives?*
>
> **Prakash Nair,** *Schools - Good for Children?*[C57]

Essential Element 2

Do your classrooms reflect your educational philosophy - your beliefs around learning and the teacher's role?

Do your virtual learning spaces support the pedagogical vision?

The learning space strongly influences how the teacher interacts with learners and learners' understanding of how they are expected to work and interact with both the teacher and the other students. For example, a traditional classroom setup in which all learners face forward, eyes focused on the teacher, clearly supports lecture-based learning, not collaborative or even self-directed learning.

Questions & Actions

Image Credit: © Jestico + Whiles

> **Look at your pedagogical vision and determine what changes you will need to make to support your vision.**

Essential Element 2

1. Begin by researching information and ideas in the book **The Language of School Design** by Prakash Nair and Randall Fielding or at the **Designshare**^C45 website. Or look at the resources at **The Third Teacher**^C46 or read some stories at **The Third Teacher+**.^C47

> *Some few years ago I was looking around at the school supply stores in the city, trying to find desks and chairs which seemed thoroughly suitable from all points of view - artistic, hygienic, and educational – to the needs of the children.*
>
> *We had a good deal of difficulty in finding what we needed, and finally one dealer, more intelligent than the rest, made this remark: "I'm afraid we have not what you want. You want something at which the children may work; these are all for listening.*
>
> **John Dewey,** *The School and Society* and *The Child and the Curriculum*^C48

Do your learning spaces cater to different learning modalities? Are the spaces flexible and engaging?

1. Watch and discuss the following videos. What examples can you see in these videos that might better reflect the sorts of learning environments that are consistent with your expectations about the conditions under which your students will learn best?

   - **Hellerup School**^C49
   - **Physical Learning Environments**^C50
   - **School 2.0 - Designing Tomorrow's Schools**^C51

2. Workplaces, too, are rethinking their spaces and their experiences can provide interesting and valuable insights on the impact of design.

   Read and discuss the following article that describes Uber's move from an open office concept to one that is designed around neighborhoods: **Goodbye open office, hello office neighborhoods**.^C52

Essential Element 2

3. Determine what changes can be made immediately, and which ones require more time or costs.

Keep in mind what research confirms: students conform to physical spaces.

Remember, not all design changes may require large capital investments.

We shape our environment and then it shapes us.

— **Winston Churchill**

What do your students think of the design?

1. Develop a simple survey to solicit student opinions and ideas.

   The students' perspective in the classroom is very different from that of teachers and school leaders and will help you develop a complete picture. They have an important stake in helping design learning spaces that best support their learning.

2. Include student representation on your planning team.

Learning spaces refers to more than just classrooms.

1. Consider how you use other spaces in your school, such as hallways, specialized rooms, the school library. How can their use be rethought to support learning?

2. Watch and discuss **Transforming Libraries**,[C53] in which Superintendent Pam Moran, Albermarle, VA, USA, talks about the changes they have made and the reasons behind them.

Essential Element 2

What is Architecture's Role in a Revolution?

— SAAL Exhibit, Canadian Center for Architecture, 2015

What is a digital classroom and learning space? What does it make possible?

1. Watch the following videos and discuss how children are learning differently in these environments. What new learning opportunities are possible?
   - **From Chalk to Screen: Six Months with the Immersive Classroom**[C54]
   - **Future Classroom Lab - Learning zones**[C55]

   How does the role of the teacher shift within these environments? What beliefs about how children learn are reflected in these environments? What digital tools are available?

2. Determine what virtual learning spaces you will implement and what tools you will use to create virtual learning spaces that reflect your pedagogical vision.

   Since learning happens in virtual as well as physical spaces, a holistic approach to understanding learning spaces helps create a natural movement within and between spaces. Software and hardware choices are an integral element and the important gateway between the physical and virtual components of the learning environment.

   Research software options for the virtual learning environment. Select the software that best reflects your pedagogical vision. At a minimum, the virtual learning spaces should include:

   - Communication and collaboration functionality.
   - Support for diverse student needs.
   - Spaces in which both teachers and students can share resources.
   - Ability to be available to students not only when in class, but whenever needed.

3. Plan to provide professional development on how to create the virtual learning space so that it is a natural part of the total learning environment.

Essential Element 2

Technology provides new opportunities to create equitable access for all students. Every classroom has students who can benefit from using technology designed for accessibility for all, ranging from those with special needs to those with mild impairments or learning differences.

1. Read **Creating a "Least Restrictive Environment" with Mobile Devices**.[C64]

How can the use of mobile devices benefit all students, not only those with disabilities or special needs?

2. Take some time to review the ideas and resources found at **Accessibility in the Classroom**.[C56]

Discuss the experiences of schools around the world in using digital devices to provide all students with equal access to the learning opportunities 1:1 makes possible.

The examples that often most stand out and illustrate the transformative potential of technology are those that use accessible technology integration to empower and enrich the world of students that otherwise might have had an extremely difficult time communicating, collaborating, or socializing with their peers."

— **Anthony Salcito**,
Accessibility: A Guide for Educators[C65]

Essential Element 2

PLANNING OUTCOMES

As a result of your research, readings and reflections…
- Create a list of essential and recommended changes to the physical learning environment.
- List your selected virtual learning space tool(s) and describe why you are recommending these.

Learning will take place here…

….in order to…..

Our physical learning spaces will reflect our beliefs about how our students learn in the following ways…

Our learning environment will include…

…….so that…..

Learning will take place at these times…

…..which means…..

Our virtual learning spaces will help us achieve our vision by….

Additional Resources Used

Essential Element 3 - Cultures of Change

PRINCIPLE: Building sustainable change requires strategic thinking to impact the organizational culture of the school.

> *Achieving, successful leaders know what is important. And they always are reminding people of what's important. It's not enough to just have vision; it's not enough to just know your objectives. Instead, a leader must create an environment where people are aware of why they are there.*
>
> *You can see this even in a university. Occasionally when we will gather in the faculty club and talk, someone will say - always in jest to conceal the basic truth - "wouldn't this be a good place to be if only there weren't students around." Well, why are we there? Why are we administrating school systems and universities?*
>
> *Primarily we are there to help educate students to be successful in life. Everything else is a cost factor or a commentary.*
>
> — **Warren Bennis,** *Introducing Change* [D1]

How might you create the best conditions for this inevitable shift?

The vision and goals have been set, the learning environment designed. In **Essential Element 3**, it is time to turn the vision into reality, theory to practice, by developing and initiating strategies in four key areas:

- Building a Change Culture
- Professional Learning
- Finance
- Communication and Policy Development

The process of shifting people's beliefs and attitudes about what is possible is a huge challenge to schools. Before any concrete implementation can take place, initiative leaders need to outline and begin to implement strategies for change across the following areas or else they will not be able to achieve their goals and vision.

- Stakeholder and community buy-in.
    - How will you communicate your vision and plans for implementation so that it can be owned and well-articulated by key members of your school community?
- Teachers' professional growth to support your initiative.
    - What professional learning strategies do you need to consider?
- Equity.
    - Core to this initiative is the belief that ALL students should have equal access to their own device. How will that be best achieved?
- Sustainable financial support.
    - How will you ensure that any funding support is not limited or short-term but can be sustained over the long term?
- Clear communication.
    - What communication strategies will be necessary to ensure all stakeholders are kept well informed?
- Policy clarity and alignment to learning goals.
    - Who will ensure that policies align to learning goals and how will they do that?

This strategic planning involves not only district and school leaders, including principals, but the Director of Curriculum and Instruction, the CFO, the Director of Professional Development, and the Director of Communication.

Project Team Members • District and/or school leaders, including principals • Teacher leaders • CFO • Director of Communication • Director of Curriculum and Instruction • Director of Professional Development

Essential Element 3

Build a Change Culture

"Time for Lesson 1 in our new language."

> What's needed is a real-time, socially constructed approach to change, so that the leader's job isn't to design a change program but to build a change platform—one that allows anyone to initiate change, recruit confederates, suggest solutions, and launch experiments.
>
> **Gary Hamel and Michele Zanini**
> *Build a Change Platform Not a Change Program*[D2]

Essential Element 3

Change doesn't just happen, nor does it always move in the right direction. Change without growth is just mucking around in the same mud. If you've articulated a vision and goals, you have a direction. Now you need to move all stakeholders down the path to this vision.

What is most effective is to build a whole school culture that not only talks about change, but takes steps in a positive direction to achieve it. This doesn't happen by changing one aspect of the environment or program, but by taking a systemic view and recognizing the interconnectedness of all components. Change doesn't happen overnight, or in only a part of the building. It isn't about one or two people in the system and it doesn't happen automatically when you hand devices to each student. It's an ongoing process that must be supported every step of the way.

> *I would suggest that one reason education reform has not worked is that it almost always treats these dimensions as separate and tries to reform one or another—the choice depending on who is doing the reforming. Curriculum reformers try to put new curriculum in an otherwise unchanged system but ignore the fact that the old curriculum really suits the system and reverts to type as soon as the reformers turn their backs. Similarly, when reformers introduce new forms of management of the old approach to knowledge and learning, the system quickly snaps back to its state of equilibrium. And, perhaps most dramatically from the point of view of people in this room, the same kind of process undermines any attempt to change education by putting a lot of computers into otherwise unchanged schools.*
>
> — **Seymour Papert,** *Perestroika and Epistemological Politics*[D3]

Essential Element 3

Questions & Actions

Explore a range of strategies for change and what they entail.

1. **Developing a culture for change means exploring and utilising some key levers and providing the time and support necessary. Read Learning to Lead Change: Building System Capacity.**[D4]

   Discuss the levers of change and the recommended strategies, especially in terms of barriers to change. Analyze what you see as your barriers to change and begin to outline how you will overcome these.

Essential Element 3

> *Leadership is the art of giving people a platform for spreading ideas that work.*
>
> — **Seth Godin** [C41]

2. Do you agree with the following statement?

> *The reality is that today's schools were simply never designed to change proactively and deeply —they were built for discipline and efficiency, enforced through hierarchy and routinization.*
>
> — **Gary Hamel and Michele Zanini** [D2]

What are the structures and policies within your school that sustain a legacy culture, and what can you do to change those?

3. Read **Enabling Transformation With Strategic Planning, Organizational Capacity and Sustainability.** [D5] Discuss what behavioral change strategies are and how you could go about implementing these. Consider the questions included in the article.

> *Change is built on behaviour. School leaders need to be thinking in terms of behaviour change strategies.*
>
> — **Ben Jensen,** *Enabling Transformation With Strategic Planning, Organizational Capacity and Sustainability*

Essential Element 3

4. Explore one or two change process strategies, such as **Lean Six Sigma**,[D6] **ADKAR**,[D7] or Kotter's **8 Step Process for Leading Change**.[D8]

These provide ideas on leading and supporting change, including moving from a hierarchical model of management to a networked model.

It is interesting to note that setting up a Guiding Coalition offers schools a chance to build a more distributed model of leadership which brings in a much more diverse range of abilities and personnel than is traditionally the case.

Building a Guiding Coalition

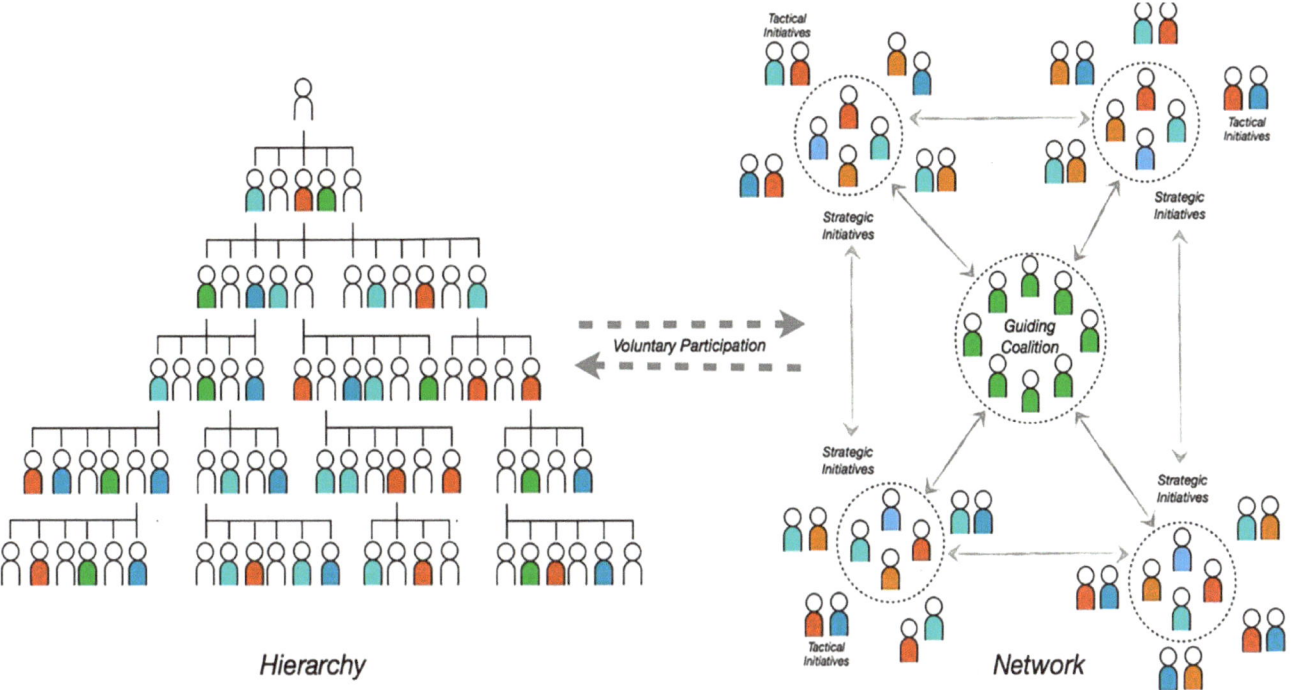

Image from *Accelerate: Building Strategic Agility for a Faster-Moving World*, John Kotter

Essential Element 3

1. Determine what the strategic priorities arising from your goals are.

> *The challenge is that educational training institutions might have talked about new paradigms and pedagogy, but are still predominantly teaching that from a paradigm of separation – separate teachers, working in separate spaces with separate programs, assessed and appraised separately. Teachers need to approach working with colleagues as if moving into partnership with others. Once the transactions of communal living have been negotiated there is a fantastic opportunity to create shared spaces, shared programs, shared learning, shared success, and shared problem solving.*
>
> **Stephen Harris**, *Executive Director, Sydney Centre for Innovation in Learning, Australia* [C43]

 Two key questions are: Who are the people who will help create a shift within the organization? What are their expectations?

1. Determine who will be the people who will have the most impact in shifting the attitudes and beliefs of your staff.

These won't necessarily be the technology adventurers who are very comfortable already with technology. Nor will they be the 'unwise', those who see no purpose to the use of technology in education. Rather, focus on those people who are great educators who are still undecided about 1:1 but willing to explore what it makes possible. They are the **Transformers**, the large group in the middle who have the potential to be real change agents. These are the people who will be instrumental in convincing other, more cautious staff members.

Essential Element 3

Who are the **Transformers** in your school?

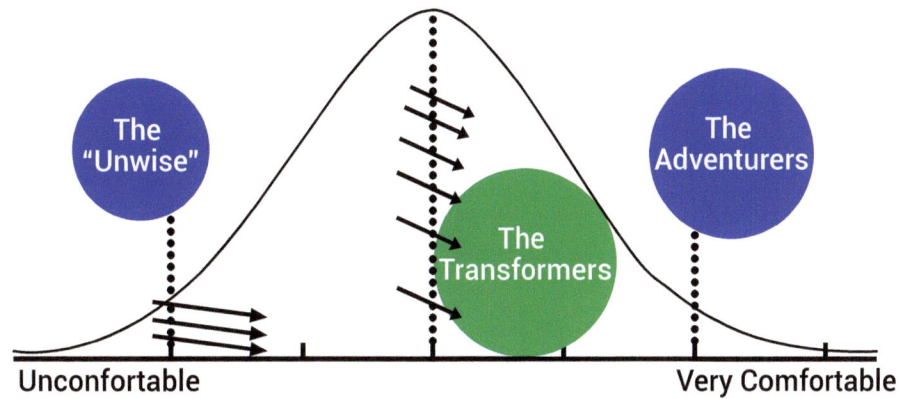

Technology Comfort Level

2. Provide your Transformers with time - the time to learn about how ubiquitous access is being used to create new learning opportunities for their students.

Remember, it is **always about the technology increasing a teacher's pedagogical capacity.** That must be the focus and the driver for this group. How else can you support them?

How and at what pace can change be made?

> It seemed the height of unreason to suppose the earth was round instead of flat, or that it moved instead of the sun, or that objects required a force to stop them when in motion, instead of a force to keep them moving, and so on.
>
> **It is only afterward that a new idea seems reasonable. To begin with, it usually seems unreasonable.**
>
> — **Isaac Asimov**, *Isaac Asimov Asks, "How Do People Get New Ideas?"* [D9]

Essential Element 3

1. Watch the video **Lead the Shift: Initiate Change**.^{D10}

   In this video, education leaders from four countries describe some of the strategies they use to help bring about a shift in learning and teaching. Use these ideas to begin outlining your change strategy, including:

   - What you perceive as the role of school leaders and teachers in the process.
   - How you will use professional development and teacher collaboration to help bring about change.
   - What system-wide support you'll put in place to bring about this shift.

2. Read and discuss the posts at **Change Management & Culture of Innovation**.^{D11}

3. Provide research evidence about the value of the shifts you envision. Explore additional strategies for getting school faculty to understand and then take ownership of the vision and goals. What are essential elements of building teacher capacity?

4. Continue work on aligning expectations and goals throughout the system. Outline how you will do this. Systemic alignment to the goal of shifting to innovative teaching practices is required in order to provide ongoing support and development.

> **Incremental change or fundamental change?**
> This important question is rarely discussed, but it is important to create a shared belief around what each means.

1. Discuss the following quotation. Do you agree with Professor Papert's ideas about incremental change? Why or why not?

Essential Element 3

> *Incremental change can be self-defeating; it's not a step on the way to the big change. A silly example: suppose that the inventor of the refrigerator found that the only way to persuade people to buy them would be to make a refrigerator that could drop the temperature by just one degree. Now that thing would be no use as a refrigerator, it would be a kind of step towards a real refrigerator. If you distributed these around people would develop ways of using them, they'd use them as storage boxes, they'd use them for all sorts of things because people are ingenious beings and they try to use what they've got. So, there'd come about a refrigerator culture based on ways to use refrigerators for purposes that had nothing to do with what we know refrigerators are good for... this is what's happened to computers in schools. They're being used in ways that have nothing to do with the potential of the computer to allow the possibility of a radically different way of learning.*
>
> **Seymour Papert,**
> *Looking at School Through School-Colored Spectacles* [D12]

2. Fundamental change or transformation - without a clear understanding of what it means, it's difficult to determine if this is what you envision for your schools. Change happens all the time, so it is essential you clarify if change and transformation are the same, and why or why not. Discuss what you believe is the best approach for your school - fundamental or incremental change?

   As you do, clarify what you talk about when you talk about change and transformation and what it would look like. This is an important conversation if you want to ensure there is a shared understanding and that goals and actions are aligned throughout your school system. Come to a consensus within your team.

3. Discuss the following paper from the Boston Consulting Group: **An Atlas of Strategy Traps**.[D13] Consider how this applies to schools, in general, and your school/district strategies in particular.

PLANNING OUTCOMES

As a result of your research, readings and reflections…..
- Describe your strategy for building a culture that supports change and innovation.
- Develop a proposed timeline for this work.

Strategies to Create and Support a Change Culture

Key people to help lead the shift

Conditions for Success

Timeline

Essential Element 3

Implement Professional Learning Strategies

- Professional development as a term is a major obstacle to progress in teacher learning;
- We need to deeply appreciate the meaning of noted educator Richard Elmore's observation (2004) that improvement above all entails "learning to do the right things in the setting where you work" (School Reform from the Inside Out);
- Student learning depends on every teacher learning all the time.

Michael Fullan, *Change the terms for teacher learning*[D14]

Essential Element 3

What are the essential elements that underpin the development of a contemporary learning culture?

No one denies the value of professional learning for their educators. Much time and money is set aside annually to implement programs that may have some benefits for the individual participant, but may not be sufficient to shift practice, nor beneficial to the school community as a whole. Much of what is called 'training' or 'professional development' is sporadic, skill-based, non-contextualized and isolated from actual practice, unevenly distributed, and insufficient. Too often there is no requirement for the participating educators to intentionally examine how they will use ideas to which they have been introduced to improve the design of learning in their classrooms or schools. What is more, it is even less likely they are called on to reflect on the effectiveness of any resulting new practices, without which positive shifts are less likely to be shared with the school's educator community.

Schools need to be cohesive learning communities, in which all learners and teachers alike are valued and respected and everyone is accountable for not only his or her personal effectiveness but the effectiveness of the community as a whole. Professional learning strategies for educators should be designed around the same model of learning outlined for students, namely, that educators are self-directed, connected, inquiry-based learners.

It is important that we accept that today, contemporary professional learning about every aspect of teaching, not just technology skills, should be **self-directed**. If we expect our students to be self-directed learners it is critical we walk in their shoes and model that behavior to them at all times.

No longer should teachers wait until a workshop is available for them to attend to gain skills.

No longer should there be a need for a technology expert to answer every call or question.

And no longer should professional learning be seen as a "push-something" that others deliver, but rather it must be a "pull-something" that you, as an individual, need to know and for which you are prepared to take the initiative to find answers.

This is the essential nature of continuous professional learning, and it reaches out to a diverse range of resources from websites, to white papers, from YouTube videos to webcasts, and from one-on-one discussions with a peer or coach to selective events, workshops, or even occasional conferences and meetings.

Essential Element 3

Questions & Actions

What professional learning strategies will you develop that will do all of the following:

- **Provide learning that is continuous and embedded in teaching practice (for example, peer coaching)**
- **Let educators participate in ongoing, meaningful action research to explore their teaching practices**
- **Boost confidence, build competence, and foster commitment among teachers?**

Essential Element 3

1. View **Professional Learning – A Critical Priority**.^{D15}

In this video, six professional development specialists and school leaders explain a number of professional learning strategies. Various individual strategies are described including breakout sessions, observing other teachers, action research, and coaching.

Discuss what the benefits, implications, and challenges of each strategy are. Have your team spend some time discussing each idea in detail. What elements are intrinsic to each strategy? What would you need to do to implement each strategy in terms of planning, scheduling, and support?

2. Read and discuss the following two research reports:

- **Professional Learning in Effective Schools: The 7 Principles of Highly Effective Professional Learning** [D16]
- **Learning Communities and Support** [D17]

How will you apply research results and ideas to your professional learning plan? Outline how you can build and support an effective community of practice within your school.

> *Creating cultures of learning in schools starts with the leaders in those schools acting as learners first and foremost. That means more than just talking about the latest conference or workshop, or the latest class at the local (or online) university. Instead, it means constantly articulating and sharing personal questions that you're interested in answering, and then sharing your process for answering those questions transparently.*
>
> **Will Richardson**
> *Eight New Attributes of Modern Educational Leaders* [B34]

Essential Element 3

1. Research strategies used by other schools, such as:

   - ePotential ICT Capabilities Resource [D18]
   - Smart Classrooms Bytes--eLearning for Smart Classrooms [D19]
   - 100 Model Classrooms [D20]

   List the benefits, challenges, timelines, and support required for each strategy. Include strategies such as coaching, peer coaching, and action research.

2. Review frameworks such as **TPACK** [D21] and the **SAMR** [D22] model, which can be used to scaffold technology use and teaching in the classroom. Determine if you will use these or similar frameworks as part of your professional learning strategy.

3. Read blog posts at **Building Teacher Capacity** [D23] and **Professional Learning Communities**. [D24] What changes would you need to make to support the development and work of a PLC? Outline the first steps.

4. Structured collaboration among educators is the key factor in developing innovative teaching practices. This is often a big shift for many teachers. Outline steps to make teaching more collaborative, taking into consideration scheduling and time requirements. What would this collaboration look like?

Essential Element 3

How will you support these professional learning strategies? What are your expectations in terms of how quickly you'll see change? How will you communicate the idea that change is expected and inevitable?

1. **Make sure your expectations for how long it takes to see change are realistic.**

   Remember that not everyone will be at the same point in terms of comfort level or technical knowledge, so plan accordingly. One-size will not fit all. Analyze your goals to see if your expectations in terms of change are reflected in your goals.

A Technology Integration Coach is a key role. The person or persons fulfilling this role will need sufficient time to work with teachers both before and after students receive their devices. Main attributes of a coach are:

- Good listener, with excellent social skills.
- Enough depth and breadth of pedagogical knowledge and strategies to help teachers who are at various stages of technology integration.
- Knowledge of how to organize/structure a technology-rich classroom and awareness of relevant classroom management skills.
- Plans technology rich activities or projects with individual teachers.
- Knowledge of effective grouping strategies and able to partner with staff in developing integration opportunities.
- Knowledge of curriculum framework and how technology can support it.
- Recognized by staff as a strong teacher who will keep teachers up to date with current research on issues related to the integration of learning technologies.

Essential Element 3

1. Read and discuss the following white paper from the International Society for Technology in Education that introduces three coaching models: **Technology, Coaching, and Community Power Partners for Improved Professional Development in Primary and Secondary Education**.D25

> Peer coaching, in which teachers observe each other in the classroom and provide feedback and support focused on specific lessons or teaching practices, offers ongoing, embedded, cost-effective professional learning. Peer coaches help other colleagues integrate technology into classroom learning activities and improve lesson design by incorporating engaging strategies, such as project-based learning.

> *Teachers tell us, they're not looking for an expert. They're not looking for an answer from someone else. They're really looking for somebody that helps them **think more deeply**.*
>
> **Les Foltos,** *Insights into Peer Coaching* D26

1. Learn more about peer coaching. Read: **Successful Collaboration: What every educator can learn from coaches**.D27

2. Watch and discuss the following video: **Insights into Peer Coaching**.D28

> Action Research can be a very effective method of professional learning and means of changing practice. Action Research entails more than just research. As its name clearly states, it also means taking action by actively testing new ideas and practices and evaluating how effective they are in terms of benefiting student learning. It is important to emphasize that the evaluation stage is critical to understanding the impact of the new practice on student learning.

Essential Element 3

1. Explore processes around action research. Read and discuss **Action Research: Insight & Advice**.[D29]

2. Determine how you can develop an Action Research program and culture. Start by answering the following questions:
   - How will you ensure effective execution of this program across your entire faculty?
   - How will you provide the necessary support to allow teachers to design, implement, and evaluate their action research work?
   - What mechanisms will you put in place to ensure that the results of this research is shared across the faculty and used to positively impact learning?

> *Technology will not make a bad teacher good, but can make a good teacher great.*
>
> — **Rob Baker, Cincinnati Country Day School** [D30]

Those First Few Months

How will your teachers manage the first day, week, month of your initiative? In spite of efforts to prepare educators through professional learning programs prior to deployment, those first few months may seem overwhelming. It is important to put in place a process that helps your teachers reflect on how shifts in their practice impact learning in the classroom. Although there will be a great deal of concern about technical issues and on gaining technical skills, it is essential to maintain a strong focus on pedagogy. Although the technical issues are genuine concerns, they can be handled through well planned management of the deployment. Remember: without a solid pedagogical foundation for understanding the shifts in learning and teaching that should happen in a well thought out initiative, the chances of having any impact on achieving learning goals are greatly diminished.

Essential Element 3

The best thing a teacher can do is to set up the best conditions for each kid to learn. Once you have that, then the computer can help immeasurably. Conversely, just putting computers in the schools without creating a rich learning environment is useless -- worse than useless!

— **Alan Kay,** *The Dynabook Revisited* [D42]

Questions & Actions

How will you help your teachers maintain their focus on pedagogy? Here are a few ideas:

Essential Element 3

1. Review the **12 Principles of Contemporary Teaching and Learning** with teachers. Provide time to discuss how they will begin to put these principles into practice in their classrooms.

   Have teachers look at the curriculum to find places where they might begin to use these principles to redesign learning activities and teaching practices.

2. Read **How to Create Rich Learning Environments**.^{D31}

   How will you and your teachers identify, articulate, and collaborate around:
   - Setting up the best conditions for each student to learn?
   - Creating a rich learning environment?

3. One way to create rich learning environments is to have each teacher prepare a **Pedagogy Statement** to which they should refer throughout the school year.

   In this, they identify:
   - Their vision for 1:1 in the classroom. What is the potential for this powerful thinking, learning, doing tool and how will this affect their instructional practices and student learning opportunities?
   - Their learning theory (how people learn).
   - The actions they, the teachers, will take as a result of this theory.
   - The actions students will take in connection with this theory.

   Sample statements and the template for creating a Pedagogy Statement are included in **Part III - Tools and Resources**. Additional samples are available on the AnytimeAnywhereLearners website.[D32]

4. Have teachers develop a set of classroom routines as well as learning routines around the laptops to help make the first few weeks more successful.

Essential Element 3

5. **Outline your plans around building professional learning communities with a focus on action research.**

As mentioned earlier, two of the best ways to extend professional learning and ensure teachers have agency in their learning is to have them form Professional Learning Communities (PLCs) and engage in action research. Before this can take place, teachers will need to understand what PLCs are, what action research is, and how they can use these to help explore and change their practices to be more effective.

6. **Remember, ongoing collaboration with other teachers and observing each other's classrooms helps support the development of new teaching and learning practices. How will you support this type of collaboration?**

PLANNING OUTCOMES

As a result of your research, readings and reflections…..

- Draft an outline of your professional learning plan built around the elements you believe are intrinsic to your strategy.
- Describe how your strategy supports change and innovative teaching practice. Be explicit. Remember that professional learning should focus on the intersection between pedagogy and technology.
- Create a multi-year strategy for professional learning in order to plan the commitment in terms of time and financial support. Outline a proposed timeline.

Professional Development Strategies

Reasons These Strategies Will Achieve Vision

Conditions for Success

Timeline

Essential Element 3

Develop Funding Strategies for Equity & Sustainability

> The basic foundation on which 1-to-1 learning was established was equity and universal access. In fact, if the initial concept of 1-to-1 learning had simply been built around the idea of allowing any student fortunate enough to have a laptop at home to bring it to school, the idea would have joined the exceptionally long list of failed educational innovations. At the heart of good 1-to-1 learning is equity to ensure that all students have equal access to technology-rich experiences and simplicity to ensure that it is easy to manage and sustain. [D34]
>
> — B. Dixon and S. Tierney,
> *Bring Your Own Device to School*

Essential Element 3

Your financial strategy should focus on four simple but important principles...

- Funding should ensure **all** students can participate and access the best learning opportunities with their **own** fully-functional device.
- Funding should be structured to ensure it can be **sustained and supported indefinitely** and not be dependent on single source, time-limited funding.
- Funding must be supported by a commitment to professional learning.
- Everyone who benefits **should make some contribution**.

Questions & Actions

1:1 initiatives are never initiated with the sole purpose of increasing the amount of technology in a school. Funding strategies must be focused on realizing the vision for the program. To achieve any vision, there must be support for a range of key components.

1. List the key areas in which you will need to invest in order to achieve your vision and the best learning opportunities for all learners.

There are a number of different ownership models of devices, each requiring careful consideration and policy development.

1. Read and discuss: **Who owns the laptops and tablets used by students and teachers, and how does this affect their use?** [D33]

If you are considering BYOD, analyze whether or not this is a sound approach, financially and pedagogically. Will this approach help provide the best learning opportunities over time?

1. Read **Bring Your Own Device to School** [D34] and **BYOD in Education, A Report for Australia and New Zealand: Nine Conversations for Successful BYOD Decision Making** [D35] to evaluate both the benefits and the issues of this acquisition strategy. Keep in mind that your financial strategies must support pedagogical goals.

Essential Element 3

Pedagogical goals should never be compromised when planning a financial strategy. BYOD raises a number of difficult, and potentially costly, issues around:

- Equity
- Infrastructure and support
- Impact on pedagogical choices
- Finances

If considering this option, carefully analyze the impact in each of these areas.

> *Digital technologies are not just a workforce issue. They are deeply embedded in the ways we live, work, play, socialize, learn, and teach. Students not only need to learn how to participate in, but also how to make, the digital world they live in.*
>
> *How do we open the door to students who have been denied access to this knowledge? Schools and teachers must become the force of equity.*
>
> **Y. Kafai and J. Margolies,**
> *Why the 'coding for all' movement is more than a boutique reform* [D36]

Providing equal opportunities for all young people means more than just providing or allowing the use of a device in school. This is also not about providing school or district owned devices, but rather one that the students see as their own.

1. **Discuss the following questions when considering what the term "equal opportunities" means.**

   Are opportunities equal when:
   - Not all children have access to the internet outside as well as inside school?
   - Some have laptops and others have phones, only?
   - Not all schools within a district have adequate bandwidth?
   - Only some schools provide a program designed around the principles of modern learning?

   What else must be considered?

Essential Element 3

2. Read **Community Hotspots for Learning**[D37] to see how the Kent School District, WA, is tackling the challenge of access for all.

> What strategy will you use to finance your initiative and ensure that **all** students have equal and excellent opportunities to learn and your initiative is sustainable? Who should pay for the cost of a student device if it is to be used for both school and private use?

Financing your initiative with one-time grants or bonds may seem like a good idea, or an opportunity not to be missed, but once the grant or bond money runs out, you will be left going through the process of finding new funding or, as too often happens, reducing or eliminating core components of your initiative. This need not be the case. Consider an alternative funding strategy right from the start: **Co-contribution**.

School communities are often blinded by past practice, where students accessed *school* computers for use at *school*. This implied the school would fund 100% of the cost of the computers. But what if the students had 24/7 access to their own device both at school and at home?

Co-contribution provides a simple answer and is a very popular funding strategy that embraces the simple idea that if students are to have 24/7 access to their own devices, then research has shown they will only be using them for school work for around 20% of the time. The rest of the time it is used for personal use. It, therefore, seems reasonable to have parents and the school *co-contribute* to the cost.

The amount contributed by parents is set by the school community or board and usually ranges from 40% to 60% of the total cost of the device and associated software and bag.

(Note that in very recent times, the BYOD label has become very misleading because it implies that if students are using their *own* device, it must be one they have brought from home. As can be seen with co-contribution, this is not the only or even best option available.)

Additional Options: A further option to minimize the impact on parents is for them to make their contribution, whether monthly, quarterly or annually, through a specially created foundation or similar body directly associated with the school or district. This also provides an option to source additional funding from other sources such as foundations, corporate donations, regional, state or national contributions, all of which can contribute to the funding pool and reduce the overall cost to parents and the school.

Essential Element 3

The great benefit of co-contribution is that it represents a saving for both parents, who alternatively would have to cover 100% of the cost, and for the school, as it can now spread their funding across a larger number of students or over a longer period of time.

The co-contribution model has been popular across many countries for more than 20 years, and allows schools to share the cost of students having 24/7 access to their *own* device in a manner that is sustainable over the longer term.

One Example of a Co-contribution Model

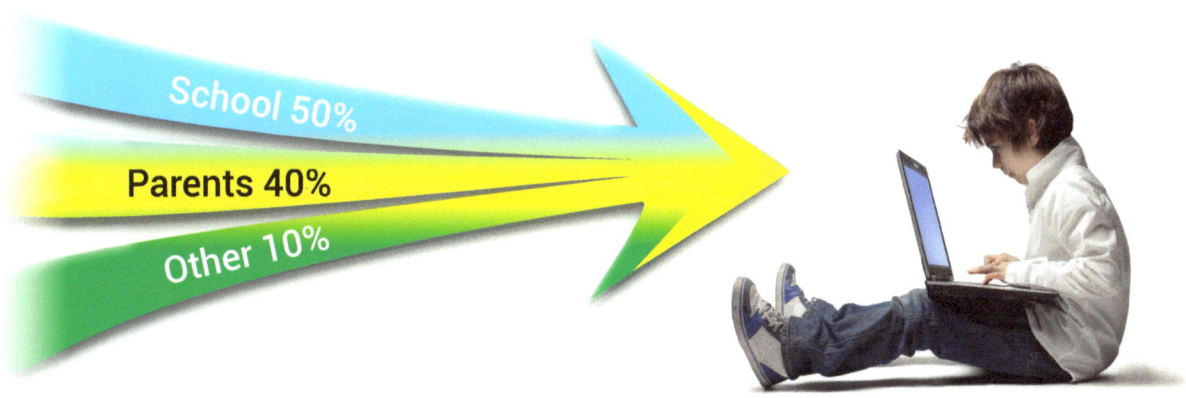

Co-contribution is a flexible funding option that is sustainable, scalable, and replicable.

1. Use the **Co-Contribution Model for 1:1 Funding Equity** template[D38] to explore some of the possibilities and develop a strategy that will provide equal and excellent access to the learning opportunities 1:1 makes possible.

2. Determine possible additional funding sources and evaluate how sustainable they are in the medium to long term.

3. If you also want to provide an option for students in some grade levels to bring their own devices from home, (for example, in senior grades), make sure you follow these guidelines:
   - Determine acceptable feature set for any personal devices that will be brought from home. The features must support the school's vision and goals.
   - Offer an option for students to participate in a co-contribution program to purchase a laptop for those who do not have, or do not want to bring, their own device.
   - Make available supplementary equity funding support for those fiscally challenged.

Essential Element 3

PLANNING OUTCOMES

As a result of your research, readings and reflections….

- Outline your funding strategy and steps to take for its implementation.
  - What are your possible sources of funding?
  - How sustainable are they in the medium to long-term?
  - What would you predict would be the likely funding mix for your initiative?
- Explain why you have chosen this strategy.
- Outline your timeline.

Our financial strategy is:

Our reasons for this strategy are:

Timeline:

Build Community Support

It is worth noting that those of us in the educational technology space draw on terms and concepts straddling numerous disciplines – psychology, learning theory, technology, and social trends (Freire/Illich-type power issues with a smattering of democracy and undertone of power and oppression thrown in) – each generally viewed to be fairly incomprehensible, but when carefully blended, is absolutely alien to the daily thinking habits of most people.

— George Siemen [C5]

What communication strategies will build the strongest support across all stakeholders within the school community?

No parent in your school community was educated in a 1:1 class or in what we would today call a technology-rich learning environment. So what expectations do your parents have of what this 1:1 initiative will make possible for their children?

It is with this context in mind that the importance of developing a long-term communication strategy should be developed.

Questions & Actions

Essential Element 3

The following table highlights the range of stakeholders for any initiative. You may think of other groups, too.

Parents	Community leaders	Teachers
Learners	Student councils	Sponsors
Teacher Unions	Destination schools	Feeder schools
Local residents	Alumni	Web community
Local businesses	Local government	School advisors
Define subsets of any of the above	Local charities	Senior Management

Dan Buckley, 2005

1. Carefully consider how involved each of these groups will be in your initiative and rank them on a scale of 1 - 7, from 1, informed, to 6, co-developed, and 7, ownership. How will you effectively communicate with each of these groups?

> *You can have brilliant ideas, but if you can't get them across, your ideas won't get you anywhere.*
> — **Lee Iacocca**

What communication strategies will create a clear, consistent message for all stakeholders?

Essential Element 3

1. **Ensure you have a common, consistent language to describe the initiative to others, both internally and to the outside world.**

   It is not always easy to explain to other stakeholders (or even agree among the educators on your team) what the various terms for 21st century skills and competencies mean, especially with so many terms and buzzwords being thrown around by the media. Even experts have problems with this.

   (You may want to read **Nonacademic Skills Are Key To Success. But What Should We Call Them?** [D39])

2. **Outline a communication plan so that all stakeholders understand your vision for learning in a technology-rich environment and the steps the school needs to take to achieve this vision.**

   Communication to all stakeholders is an ongoing, long-term process. A carefully designed communication plan not only informs the parents and community, but also makes them ardent supporters and advocates.

   What communication vehicles do you currently use?

1. **Outline how you will use these your current communication outlets to reach each stakeholder group.**

PLANNING OUTCOMES

As a result of your research, readings and reflections….
- Outline your communication plan.
- Explain how each stakeholder group will be involved.
- Prepare a communications timeline to build support and awareness.

Stakeholder Group Involvement

Communication Strategies

Timeline

Essential Element 3

Part II - Implementation

The Implementation Phase

How will you best manage the implementation of your initiative?

Once all the strategies for change are in place and are being put into action, you are ready to begin your technology implementation.

In Part II, the focus is on a number of steps that together cover the diverse range of tasks that underpin the fidelity of implementation, culminating in the actual start of the 1:1 initiative.

Implementation occurs along a number of dimensions involving several distinct groups of participants. At this point, the project team expands from the original core envisioning team to now include a number of more specialized sub-teams as various actions are clustered along specific tracks (technology infrastructure and service, teaching and learning). Their tasks are not sequential, as different project sub-teams can execute a number of tasks concurrently. Although each sub-team may have its own focus, there will not only be overlap, but a need to constantly communicate to ensure everyone's actions are aligned.

If a school or district has defined an actionable vision with unambiguous goals, and if it has developed strategies around this vision and these goals, the teams handling the implementation will have a clear path to follow.

Implementation

Project Team Members

Staff members selected for each project sub-team should be enthusiastic about the project and the changes it will bring about, and they will need to have the time to devote to it. Focus your energy on staff members who are willing to grow with the project.

Give potential team members an accurate indication of how much time they should set aside for project activities. You will need to allocate enough relief and/or substitute time to help teachers and other staff members balance their schedules. To minimize scheduling conflicts or misunderstandings, it is crucial for the team to clearly identify milestones to which all team members can commit.

The project team will oversee a range of tasks from recording serial and asset numbers to organizing laptop maintenance and repair. At the same time, the team must tackle such complex issues as:

- Gaining parent and community consensus.
- Ensuring that plans for integrating laptops into the curriculum align with learning goals and outcomes.
- Working with teachers to create professional development opportunities.
- Raising funds.

The team should include representation from all groups that will be affected by the new program:

- Faculty
- Curriculum development
- Administration
- Technology
- Academic and vocational education
- Libraries
- The school board
- Parent volunteer groups
- Business leaders
- Students

Implementation

Program Implementation Manager: The Initiative Champion

Early on, assign one of your team members to be the program implementation manager. This person will oversee the entire program implementation for the next few years—from its beginning through its growth into a mature program. For this position, look for excellent program management skills, including the ability to handle many details at once. The program implementation manager will play many roles in order to:

- Coordinate, implement, and report project milestones.
- Negotiate and coordinate with vendors.
- Design and implement an internal support and diagnostics process for the laptops.
- Coordinate service arrangements for the laptops.
- Publish progress reports and communicate with the project team.
- Facilitate staff training.
- Develop presentations and informational material for parents.

Subteams

The following list includes recommendations for various core members of subteams:

Planning (Envisioning) Team

- District and/or school leaders, including principals
- Project leader
- Program implementation manager
- Pedagogical leaders (academic directors, curriculum and instruction directors, teacher leaders)
- Chief Financial Officer
- Chief Technology Officer
- Communications Director
- Parent representative

Implementation

Technology Team
- Chief Technology Officer
- Implementation Manager
- Support Services Manager
- Diagnostic Support Technician

Pedagogical Team
- Director of Curriculum and Instruction
- Teacher leaders
- Director of Professional Development
- Student representative
- Technology Integration Coach

PLANNING OUTCOMES

As a result of your research, readings and reflections….
- List Program Manager and the members of each subteam.

Implementation

Conduct a Readiness Assessment

Project Teams
- Planning Team
- Pedagogical Team
- Technology Team

Questions & Actions

Conduct a readiness assessment to determine the school's current resource position on ICT and infrastructure, personnel, and facilities.

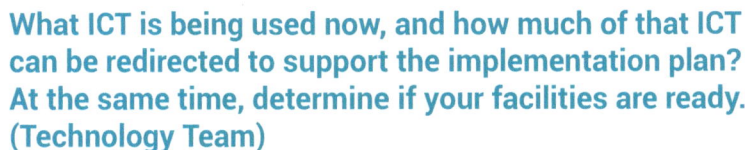

> **What ICT is being used now, and how much of that ICT can be redirected to support the implementation plan? At the same time, determine if your facilities are ready. (Technology Team)**

1. **Complete a technology audit by reviewing all existing technology and infrastructure and making recommendations based on the relative preparedness of the various components. Consider:**
   - First and of critical importance, the effectiveness and coverage of wireless across school campuses. Your target is no 'black spots' to ensure students will have access to high speed broadband at all times and in all places across the school.
   - Servers, bandwidth, hubs and routers, CD towers, and other key network components.
   - The current number of teacher laptops, student laptops, tablets and other digital devices being used for learning or for administration.
   - Printers, cameras and other peripherals.
   - Smartboards, projectors, and/or other display devices.

Implementation

2. To determine if your facilities are ready, start by answering these questions:
   - What can your school's wiring and electrical system support?
   - Will students be able to easily recharge their laptops if necessary?
   - Is there sufficient air conditioning in the rooms?
   - What about security? Where will students securely store their laptops when not in use?

3. For a detailed and in depth review and audit, use the **Microsoft Assessment and Planning Tool**[E1] to help determine your technology needs.

How ready are you in terms of teacher preparedness? (Pedagogical Team)

1. Examine how many staff have adequate skills and ICT Pedagogical competence. Consider:
   - The current amount of technology use by teachers for pedagogical purposes.
   - The current amount of technology use for administrative purposes.
   - The amount of use of technology by students for learning.
   - The amount of use of technology by students for other purposes (for example, for submitting homework).
   - How technology is being used for learning. Is it being used mainly for research? As e-book readers? For presentations? For modeling, simulating, creating projects, programming, etc? Collaborating? Communicating? Inquiry-based learning?
   - The level of support for and willingness of teachers to try innovative teaching practices.

Implementation

What are your recommendations? What are your current infrastructure priorities and how well will they scale?

1. Use the information from your readiness assessments to outline both **essential** and **important recommendations**, as well as other observations and issues worth further study.

PLANNING OUTCOMES

As a result of your research, readings and reflections….
- Describe in detail the results of your pedagogical, technology, and facility readiness assessment.
- Develop a list of recommendations for each area.

Readiness Assessment

Technology
Level of Preparedness
Essential Recommendations
Important Recommendations
Other considerations

Facilities
Level of Preparedness
Essential Recommendations
Important Recommendations
Other considerations

Pedagogy
Level of Preparedness
Essential Recommendations
Important Recommendations
Other considerations

Implementation

Consider Implementation Options and Project Plan

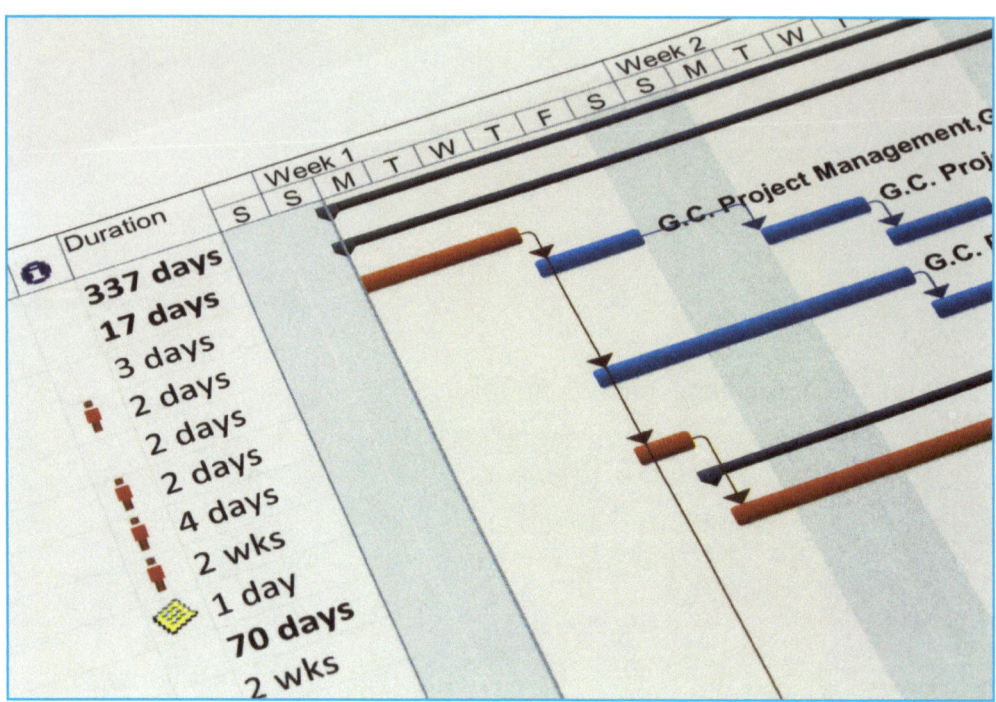

Project Team
- Planning Team

Questions & Actions

Exploring implementation options is a critical part of any 1:1 implementation and must involve discussions across a broad cross-section of stakeholders. How are you considering rolling out your program across your school(s)?

- What grade levels or sub-schools are you considering? Over what period of time?
- Which students will receive laptops in the first phase of your plan? Why? What are your reasons for making these choices?
- How will you extend the initiative in subsequent phases? How will your choices ensure the program will be sustainable over time?
- Over what length of time will the initiative be fully deployed and how long will each phase of the plan last?
- Are there other implementation options that you think should be explored?
- What technology will you need to support the initiative?

Implementation

1. Determine how you will begin and expand your initiative.

Will you begin with one class, one grade level, one class in each of several grade levels, more than one grade level at a time? Only one grade level for the whole initiative (for example, only the students in the upper grade of your school will have laptops)?

Some of these decisions will be shaped by the composition of your school. For example, if your school consists of three grade levels, you may choose to implement in the lowest grade first and expand the program one grade up each year. How and at what level you begin your initiative should be determined based on your pedagogical vision and goals.

2. Create a timetable outlining the phases.

A timeline for project implementation is typically 6–12 months for planning, but projects out 3 years. A realistic timeline has **measurable milestones** and takes into consideration that many tasks can be accomplished simultaneously.

> What are the challenges that may arise based on your phase-in plan? What are some of the challenges for the students and the teachers?

1. Explore multi-year scenarios and the impact on students and teachers at each level. For example:

- Some schools phase in laptops over several years, implementing one grade at a time as the students advance through the school. This means that in the second year, students have had one year of experience with laptops, but their teachers are just beginning to implement laptop learning.
- Some schools provide laptops for one grade only. The next year, those students who had laptops the previous year no longer have them.
- Some students have laptops in primary or middle school and then move to a secondary school that does not have a 1:1 program.

Plan how the program will move from one grade or school to the next and what concerns you'll need to address. For each scenario, consider:

- What will the response of the students be? The parents?
- What are their expectations?
- What learning goals are achieved with this approach?
- How does this impact the learning environment and teachers at the next grade?

Implementation

2. How do each of these strategies affect professional learning strategies? Review your professional learning strategies and revise them to take the timetable into consideration.

> What technology do you need to support the initiative and when must it be in place? What, if any, changes to your facility need to be completed?

1. Determine the date at which you require each component. Once your infrastructure, device, and software plan are completed, add their purchase, delivery, and installation to your timeline

PLANNING OUTCOMES

As a result of your research, readings and reflections...
- Prepare a detailed multi-year (if necessary) rollout timeline and project plan for deployment.
- Incorporate information from the infrastructure, device, and software plans, when available, into your timeline.

Rollout timeline
Grade/class rollout plan rollout:

Timetable for rollout:

Incorporate dates when infrastructure, devices, and software are required.

Implementation

Choose Your Devices, Core Tools, and Apps

> Those initiatives that have empowered students through the provision of a fully functional, personal, portable computer have derived more value and benefit because they faced fewer limitations on device use than those initiatives that focused on cost alone. These results highlight the need to focus not on cost, brand, or platform, but on functionality and the range of learning opportunities this functionality makes possible.
>
> — *A Policy Agenda for a 21st-Century Education*[A1]

Project Teams
- Pedagogical Team
- Technology Team

Implementation

Questions & Actions

Image Credit: © Luke Wroblewski

> What processes should be put in place to best inform the decisions around which device should be selected for student use? What are the key criteria that should be considered in recommending the most suitable student device?

With the help of the Pedagogical Team, carefully define which criteria are prerequisite to the learning versus what is just a preferred or optional feature.

1. Review information in the **Computing Capability Taxonomy**.[E2] What capabilities are essential to achieve the vision and goals of your initiative? Develop a list of the desired capabilities.

    Select devices based on **pedagogical goals**. Consider:

    - What capabilities will be required at various stages of your initiative.
    - Whether or not the device is strong enough, durable enough, and light enough for students.
    - If the device provides sufficient speed and memory to run required applications.

    Will you be using different devices for different age groups?

Implementation

2. Have each member of your team read at least two of the following articles and discuss how the ideas impact your device decision:
   - **Sorry But The Type Of Device Still Matters** [E3]
   - **Why Some Schools Are Selling All Their iPads** [E4]
   - **Tablets vs Laptops: Why This Is (Often) The Wrong Debate** [E5]

3. Develop a list of your device recommendations.

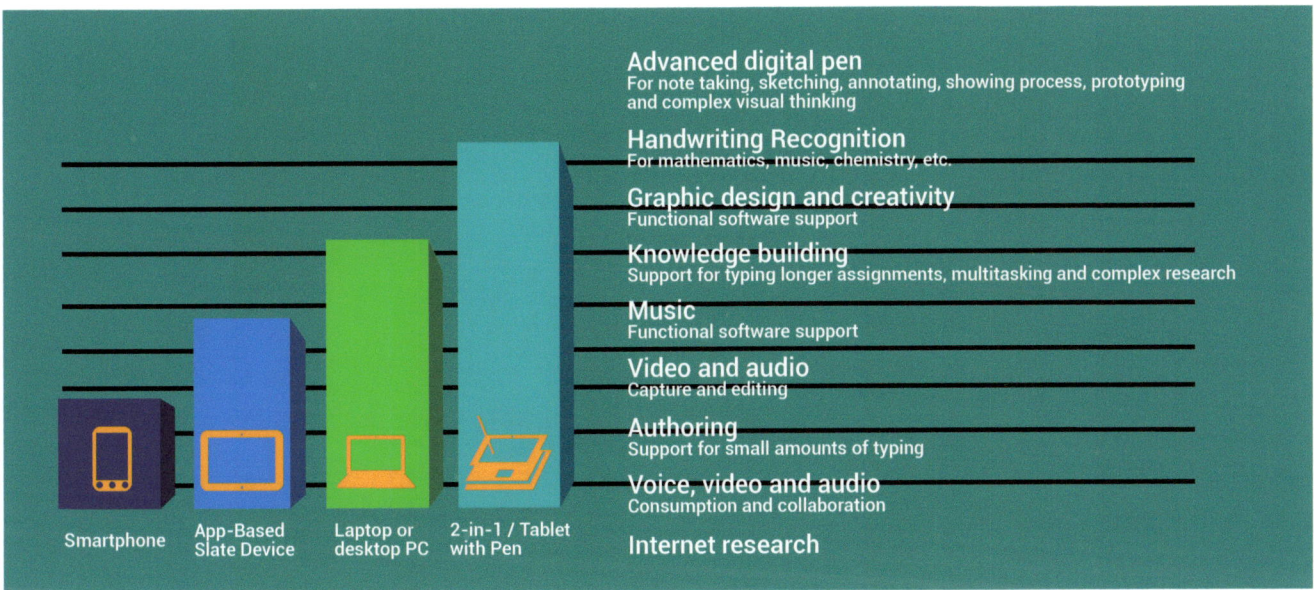

What processes should be put in place to best inform the decisions around which software will be installed on students' and teachers' laptops?

1. **BEFORE** purchasing any additional software, review your learning objectives and how you envision software will support them.

Implementation

2. Evaluate all potential software tools by asking these questions:
   - Do your selected software tools fit your pedagogical goals?
   - Are there any security, data protection, access management, or database service issues?
   - Are they accessible to all students? To learn more about accessibility issues and features, you may want to review posts at **Accessibility**.[E6]
   - If using a variety of devices, are the services and tools consistent across all of these devices?

3. You may want to review the following resources
   - **OER Commons**(Open Education Resources)[E7]
   - **LeMill**[E8]
   - **Web 2.0 Cool Tools for Schools**[E9]

Is there an easy pathway for teachers to add software they think will be of value to learning and teaching in their classrooms? How will you ensure it protects student data and privacy?

1. Outline a policy for teachers so they know what they need to do when they select software or online tools to use with their students. The process should be guided by **what will be best for the learner**.

If the process is unnecessarily cumbersome, it will impede teachers as they develop new, innovative teaching practices. On the other hand, the policy should also clarify how to protect student information.

PLANNING OUTCOMES

As a result of your research, readings and reflections….
- List criteria for device selection, which devices meet these criteria, and why.
- Create a list of software, core tools, and applications that are essential and those that are highly desired.
- Outline the process for adding new software and tools.

Key Criteria and Secondary Considerations for Device Selection:
Secondary Considerations:
Software, Core Tools, and Applications:
Policies for educators to add software and tools:

Implementation

Plan Your Infrastructure

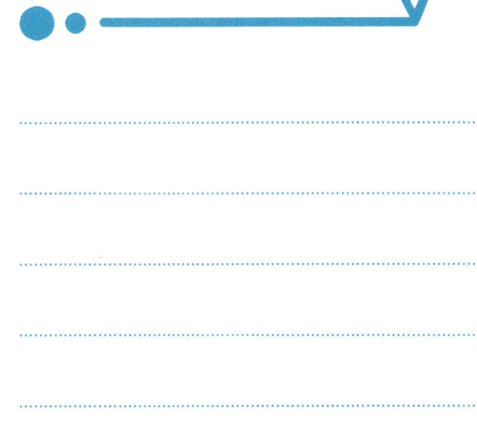

Project Team
- Technology Team

Questions & Actions

What are the key expectations staff and students have in terms of infrastructure?

Implementation

1. View and discuss the video **Infrastructure, A Strategic Asset**.[E10]

The video explains not only the main components of infrastructure, but the strategic importance of making sure it is well-designed and its processes well planned. The most common expectation of staff and students is that the **infrastructure will work whenever needed**, with no or little delay and no threat of shut-down. This should be a goal so that infrastructure supports learning rather than hinders it.

> **What are your current infrastructure priorities? How will these priorities change in light of your move to 1:1?**

1. Refer to your readiness assessment essential recommendations to determine your starting point. Moving from a 2:1 to a 1:1 program does not mean your initial infrastructure demands will only double, but, rather, it will increase much more because the frequency of use of each laptop grows. Plan accordingly.

   Infrastructure aspects to consider when designing infrastructure for 1:1 initiatives include, as a start:

   - Power sources (electricity, solar, other alternatives)
   - Internet grid options
   - Cloud strategies
   - Network design
   - Server infrastructure
   - Devices
   - Security
   - Data protection and recovery
   - Identity and access management
   - Bandwidth

2. With the help of the Pedagogical Team, carefully define what peripherals are prerequisites for achieving the learning goals versus what are just preferred or optional peripherals.

Implementation

3. Consider what services devices will need to connect to, for example: email, social media, learning management systems, e-books, etc.

4. Anticipate future needs. Build in enough flexibility to support new, innovative learning opportunities and teaching practices. Everything must continue to work.

5. Answer in detail the questions listed on the page **Infrastructure Questions** in **Part III - Tools and Resources**, taking into consideration your notes for the actions listed above.

PLANNING OUTCOMES

As a result of your research, readings and reflections...
- Prepare a detailed infrastructure plan based on current and expected future needs.
- Create a timeline for the build out of your infrastructure.

Essential Requirements:
Additional recommendations:
Other Considerations:
Build out timetable (to be incorporated into rollout plan)

Implementation

Prepare the Budget

Project Team
- Planning Team
- Technology Team

Questions & Actions

What are your budgeting priorities? (Planning Team with input from Technology Team)

1. Review your current technology related expenditures, including:
   - Infrastructure improvements
   - Hardware and software
   - Communications costs
   - Support
   - System maintenance and upgrading
   - Professional learning

Implementation

2. Prepare your budget using the information from your review (from the timetable and project plan), taking into consideration the results of the readiness assessment, and following the goals of the financial strategy.

   The budget should be multi-year and cover all aspects of the implementation, including adequate time and opportunities for professional learning. It should provide enough flexibility to cover changes due to new options for infrastructure, hardware, and software or shifts in pricing.

3. You may want to review and use the budgeting spreadsheet from GESCI (Global e-Schools and Communities Initiative), **Deploying ICT: a practical budgeting tool based on Total Costs of Ownership.**[E11]

4. Analyze how your budgeting priorities may change over time.

PLANNING OUTCOMES

As a result of your research, readings and reflections...
- Outline a multi-year budget

Multi-year budget plan

Implementation

Establish Critical Partnerships

Project Team
- Planning Team
- Technology Team
- Pedagogical Team

Questions & Actions

What partnerships should be considered to build capacity beyond immediate school resources?

Implementation

1. Explore partnerships across the range of initiative needs and available opportunities, including:
   - Infrastructure and support
   - Professional learning
   - Student learning
   - Telecommunication needs

2. Engage suppliers that have a vested interest in ensuring the program works, devices are maintained, and students have a reliable 1:1 experience.

   A decision to engage a supplier should not be made purely on price. Poor repair policies and turnaround times will negatively impact learning in your school.

3. Explore partnership opportunities and communities that support school leaders, educators, and students, for example, organizations such as ISTE, ACEL (Australia), European SchoolNet, etc.

4. Look for partnerships with other schools that can help build teacher capacity as educators from around the world share ideas on learning and teaching in a 1:1 environment.

> **How can a school best evaluate the benefit of these partnerships?**

1. Outline your goals for all partnerships. Consider how different partnerships may benefit your school to see which may best serve your needs.

> **Your supplier is one of your key partners. Defining Key Performance Indicators(KPIs) provides you with a consistent measuring stick by which to evaluate this partnership. The success of your initiative may be imperiled if these KPIs cannot be met.**

Implementation

1. Start by considering the following **6 essentials of any successful 1:1 channel partnership** and how you will use these to define your KPIs:
   - This is to be a long-term relationship.
   - Both parties must share benefit and risk.
   - This is not a traditional PC purchase relationship. It is **not** first and foremost about price.
   - The relationship should be open and transparent for both parties.
   - Expectations on both sides should be clearly documented.
   - The relationship should be built around shared goals.

2. Determine what the KPI criteria for your partnership with the supplier are and how often you will review this partnership.

   How many spare devices will they supply? What is your maximum turnaround time for repairs? What are the provider's policies for faulty devices?

3. Once the 1:1 initiative begins, keep track of any repairs and turn-around times to ensure the provider is meeting your performance measurement criteria.

   Use this information at predetermined review times to determine if your supplier is meeting your needs.

PLANNING OUTCOMES

As a result of your research, readings and reflections….

- List your Key Performance Indicators (KPIs) and KPI services review policy.
- List service provider partners, partnership opportunities with other schools and educators, and partnership opportunities with industry and community organizations that you will pursue.

Technology Providers:

KPIs and Review Policy:

Learning and Teaching Partnerships:

Community and Private Partnerships

Additional Resources Used

Implementation

Manage Support Services

Project Team
- Technology Team

Questions & Actions

> Why is it critical to ensure the use of the technology is seamless for both students and teachers?

1. View **Manage Support Services**.^{E12} Discuss what happens when a student's laptop is damaged and the potential impact a non-working laptop may have on learning and teaching in the classroom.

Implementation

What will take the device out of students' hands and who is responsible for service? What may the failures be and how many can you expect? Data collected from other 1:1 schools will help you predict the level of support a 1:1 initiative will require.

Consider the following:

Supplier responsibility

- Hardware failures and warranty repairs
- Accidental damage (insurance)

School / Student responsibility

- Software problems, such as viruses, connectivity issues, etc.
- Non-hardware connection problems

More specifically, there are two types of repairs:

A. Hardware Failures (25% of total visits)

These are 'break/fix' which require authorized service from a service partner. The following figures should be taken as worst case scenarios, but they allow for conservative estimates on the manpower and resourcing required for support services.

Hardware failures requiring service:
- Warranty 75%
- Non Warranty / Damage (insurance) 25%

Year	Expected failures (approximate)
1	40% of devices
2	100% of devices
3	120% of devices

Using this device estimated service rate (above) and the service partner's turnaround estimate, it is a simple calculation to estimate the number of 'loaner' devices that will be needed to ensure that **no** student is without a device if his or hers has to be serviced.

Implementation

B. Minor Software/Connectivity/Virus Problems

These should be handled by a student managed Help Desk, which should be first line support and have a fully trained student support team and well documented support processes that will provide data on service effectiveness. This will help the school measure whether service partners are meetings KPIs as part of their Service Level Agreements (SLA).

1. Consider this data to determine why there will be failures, what these failures might be, and how many you can expect.

2. Use the **AALF Effective 1 to 1 Service Models** spreadsheet,[E13] based on data collected from 1:1 schools around the world, to help determine your support parameters in terms of critical service indicators.

This will help you calculate the percentage of spares you need, the number and rate of repairs you can expect over a several year period, as well as the type and number of parts to keep on hand vs. those supplied as needed by your provider.

Where will support tasks be handled in the school and district and who will handle them?

1. Determine what repairs will be handled on-site and what repairs will be sent to your provider.

2. Determine what are the criteria for selecting a physical location for onsite support (size, location, any other requirements) and allocate a location.

3. Ensure you have enough staff to support the devices when they arrive.

What happens when a device is brought in for repair?

1. Create a process for logging repairs and other service or re-imaging support required. This log allows schools to collect data on the types of problems reported.

Implementation

> Students are power users, so schools need to be prepared for a certain percentage of laptops to always be away for service.

1. Determine the size of your loaner pool, as well as swap-out batteries, cables, and other easy to stock and replace components.

> Will students have a role in support services?

1. Learn more about student support roles by reading **Student Support for Laptop Programs**.[E14]

2. If you plan on involving students in running your support service, contact organizations like **GenYes**[E15] or **Mouse Squad**[E16] for information on how to set up student help desks.

---

from **Student Computing Support Teams: Learning Real World Skills That Go Beyond the Classroom** [E17]

by April-Hope Wareham, former Illinois Math and Science Academy (IMSA) Student

Working with and managing the Student Computing Support (SCS) team at the Illinois Math and Science Academy (IMSA) was the most beneficial part of my high school learning experience. The SCS team was around 25-30 students who took care of the 650 10th-12th grade students' computers.

You see, a student can only learn so much sitting behind a desk and doing repetitive homework problems. An A in an AP computer science class means nothing if you don't have the experience of an annoyed IT customer knocking on your door at eleven at night with no knowledge of how her computer "just stopped working".

An A in a business class doesn't teach you how to stay up until one in the morning online with a co-manager, hashing out the details of which team members are going to work when, who can work with whom, who has what skills and who needs a bit of help, either. Never before joining SCS did I realize the importance of teamwork, and the proper role of leadership.

What preventative measures will you take to reduce damage? How will students know what to do in case of damage and what their responsibilities are?

1. Prepare information for students on laptop care and safe usage.

2. Communicate to students, teachers, and parents what the support service policies are and what the expectations are in terms of normal wear and tear. Have this ready before any devices are distributed.

PLANNING OUTCOMES

As a result of your research, readings and reflections….
- Outline your support services plan, including all the information you have gathered above.
- Define your policies and all policy protocols.

Support Services Plan (in detail):
Staff:
Repair log-in procedure:
Support Services Location:
Support Service Policies:

Implementation

Create Effective Policies

Kent School District's technology policy can be summed up in one sentence. All use of the system must support education.[D40]

— **Kent School District, WA, USA**

Effective policies not only support your initiative, but help achieve learning goals. When establishing policies, consider what types of policies will ensure the devices are in students' hands whenever needed.

The development of a comprehensive policy document, which ultimately might be simply called the *Laptop Handbook*, is a critical step in the 1:1 journey. It sets out in detail the expectations of the school in regard to how each device will be used and cared for.

The policy document should be developed by a broad cross-section of stakeholders who need to ensure there are no 'gray areas' around use. It must cover every aspect, from use of games and personal software to backup and access to social media.

Implementation

Questions & Actions

What are the areas around use of the technology that will require policy guidelines?

1. Use the **AALF Policy Decisions** in **Part III - Tools and Resources** as a guide for outlining important areas of focus when developing your policy questions within these three key areas:
   - Effective implementation
   - Equity and scalability
   - Sustainability across all dimensions

What process will allow you to develop the most effective policies?

1. Include students on your policy decision team and discuss your policies with a range of audiences, getting input from relevant members of the staff and school community.

2. All policies must support student learning and be aligned with and support your vision and pedagogical goals. This should be your yardstick against which all policies should be evaluated.

3. Although your team will set up policies, discuss and note how you will ensure the following guidelines are kept in mind:
   - All assumptions and concerns around use are considered and the implications fully documented and shared widely.
   - Policies are determined and answers prepared **before** any parent nights and the start of the 1:1 initiative.

4. Determine how often policies will be updated and reviewed and, if necessary, changed.

Implementation

How will policies be enforced?

1. **Clearly outline what the consequences of not following a policy will be. All assumptions and concerns around use should be considered and the implications fully documented and shared widely.**

   Make sure consequences are not divorced from pedagogical needs and goals. (For example, a consequence that deprives students of their devices - devices considered to be integral to learning - contradicts beliefs that the device is integral to achieving learning outcomes.)

2. **Include information about these policies in a Responsible Use Policy document. Make sure all students and their parents have a copy of this document and understand the policies and consequences.**

3. **How will you address new policy needs? Determine the process for responding to new policy needs and enforcement. Policies should never be improvised.**

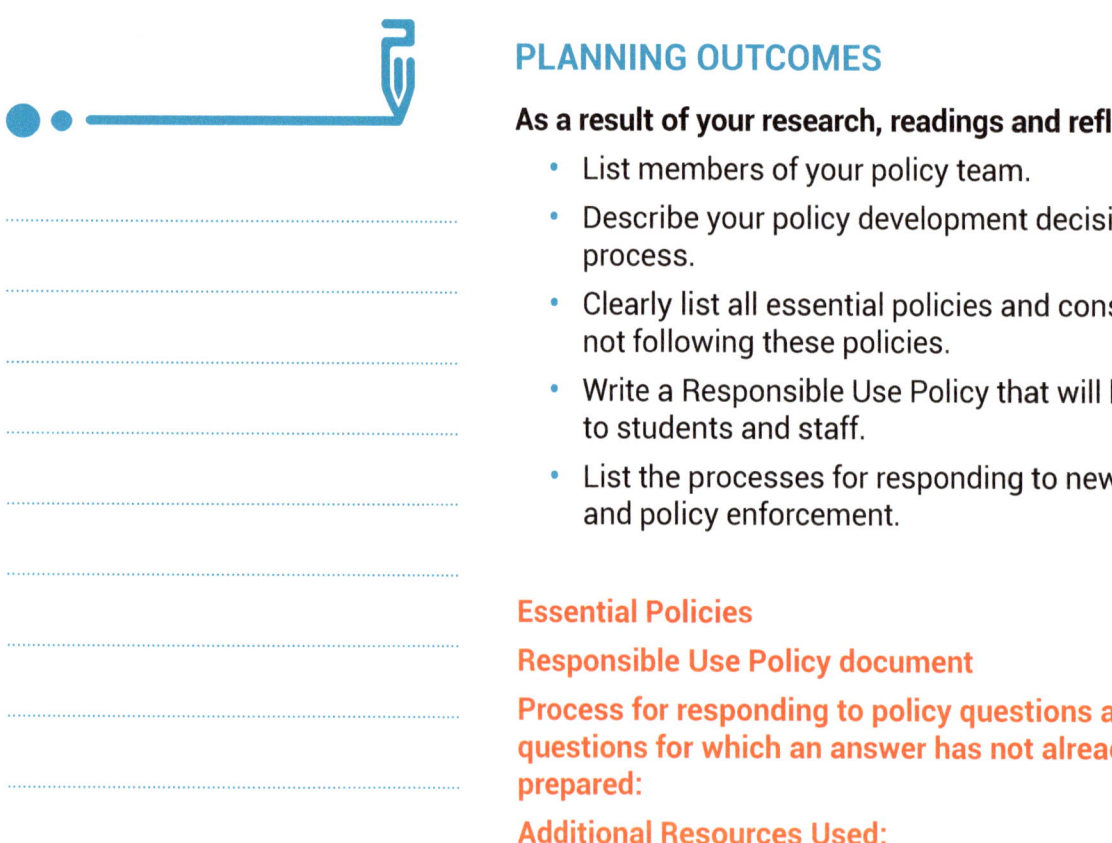

PLANNING OUTCOMES

As a result of your research, readings and reflections...

- List members of your policy team.
- Describe your policy development decision making process.
- Clearly list all essential policies and consequences of not following these policies.
- Write a Responsible Use Policy that will be distributed to students and staff.
- List the processes for responding to new policy needs and policy enforcement.

Essential Policies

Responsible Use Policy document

Process for responding to policy questions and other questions for which an answer has not already been prepared:

Additional Resources Used:

Implementation

Liaise with Parents and Community

> The two words 'information' and 'communication' are often used interchangeably, but they signify quite different things. Information is giving out; communication is getting through.
>
> — **Sydney Harris,** *Strictly Personal* [E18]

Project Teams
- Planning Team
- Pedagogical Team

Expand your communication plan to ensure a genuinely shared understanding of both the vision and reality of its implementation. Parents, in particular, will have many questions you must be ready to answer.

Implementation

Questions & Actions

> Review your communication plan to ensure all stakeholders understand the benefits and value of your 1:1 initiative. Remember, your communication plan should be ongoing, with regular releases and information targeted to each stakeholder group.

1. **Begin to communicate your plans and goals before devices are distributed. Carefully consider the timing implications in releasing information.**

   For example, a 1:1 program may save a parent from buying a home computer for a holiday present.

   Plan parent meetings and provide parents with time to respond to your ideas and ask their questions. Schools and districts should keep stakeholders regularly informed about planning, implementation, and success.

2. **Create a variety of opportunities for parents to learn about the program.**

3. **You may choose to provide parents with some basic training on the use and care of the devices. Include these sessions as part of your communication plan.**

4. **In the communication plan, include actions and events designed to keep parents and the community updated on how you are achieving your goals and to celebrate the success of the initiative.**

   This may include regular newsletters, student presentations, demonstrations of exciting ways in which the students are using their laptops for learning, etc.

Implementation

Are you prepared to answer most questions from parents?

1. **Prepare a FAQ for your parents that includes answers to 30-40 questions your parents may ask. Distribute it at the first parent information meetings.**

   Start by reviewing **Frequently Asked Questions About 1:1** in **Part III - Tools and Resources**, which lists 100+ questions parents, students, and staff may ask based on the experiences of 1:1 schools around the world. Although you may not be asked all these questions, it is a good idea to prepare some key answers before meeting with parents and most, if not all, answers before deploying the devices.

2. **Set out some guidelines so that staff and faculty are prepared for parent, student, and media questions.**

3. **How will you address questions to which you do not already have answers? Outline guidelines on how to respond to these questions.**

4. **Prepare a document with all your answers and insert it in the official initiative plan. Make sure this is available to all staff members and school leaders at all times.**

PLANNING OUTCOMES

As a result of your research, readings and reflections...

- Revise your communication plan to ensure you are regularly communicating with parents and the community to keep them informed and updated.
- Prepare your FAQ and make it available to staff and parents.

Communication Plan

Answers to Parents' Questions – FAQ

Guidelines for addressing questions not answered in FAQ

Schedule for Events, Demonstrations, and Celebrations of Success

Implementation

Deploy

Project Teams

- Planning Team
- Pedagogy Team
- Technology Team

Questions & Actions

How will you handle the preparation of devices?

1. Order devices early and make additional technical support available during this process. Determine how much time you will allow for possible delays in delivery. If students are required to have a certain school bag or case before devices are distributed, make sure these are scheduled to arrive prior to deployment day.

Implementation

2. Build the software image and make sure devices will be connected to internet, printers and/or other peripherals you will be using. Provide sufficient time for technical staff to apply software to all devices.

How will you handle the deployment of devices?

1. Set up formal agreements with parents and guardians about students' use of devices. Make sure these are returned prior to deploying devices.

2. Design a deployment strategy that is both efficient and ensures that students receive the correct devices as per the signed agreements with their parents or guardians. Implement a system to track serial numbers and register devices in the School Management System.

   Assign personnel to handle this job. Carefully consider the timing of the deployment to minimize disruptions to classroom practice.

3. Determine if you will provide student training immediately on deployment and what, if any, additional training (for example, keyboarding) will be incorporated into the curriculum in the initial stages of the initiative.

4. Determine what the criteria will be for determining when the laptops can be sent home. Decide if laptops will be sent home on distribution or only after a period of in-school training.

5. Remember to set up a process for students entering or leaving the school mid-term.

Implementation

PLANNING OUTCOMES

As a result of your research, readings and reflections...

- Write a formal agreement for parents and students.
- Create a timetable for device delivery, imaging, and deployment.
- Outline a deployment process, location, and device tracking system.

Formal agreement for parents and students

Timetable for device delivery and imaging

Tracking system

Deployment process, location, and timetable

Implementation

Part III - Tools and Resources

1:1 Checklist

✓	Actions	Responsible person(s)
	Vision, Mission, Goals	
	Research a range of 1:1 implementations nationally and internationally	
	Establish an ongoing record of the planning and implementation process	
	Define a clear, actionable vision statement for 1:1 implementation	
	Develop a common set of beliefs around how children learn	
	Define a mission based on the 1:1 vision	
	Clarify goals, expectations, and milestones	
	Develop a manageable general timetable for planning & implementation	
	Identify key project staff and assign tasks / responsibilities	
	Establish evaluation and monitoring process guidelines	
	Develop surveys / monitoring devices and timetable for implementation evaluation including:	
	• Parent / community satisfaction	
	• Staff / student satisfaction	
	• Effectiveness at achieving stated goals	
	Teaching and Learning for Contemporary Learners	
	Discuss and begin planning for shifts around when students learn - 24/7 learning	
	• Where students learn – formal and informal learning opportunities	

Tools and Resources

	• What students learn – rethink implications for curriculum	
	• How students learn – creating learning opportunities that tap into students' curiosity, let them be more self-directed, connected	
	Explore new forms of assessment	
	Investigate possible changes in schedule design	
	Research and define shifts in the role of teachers	
	Research and define shifts in the role of the principal and other school leaders	
	Configure learning spaces to best meet learning and teaching goals	
	• Physical (layout, furniture, classroom technology, etc.)	
	• Virtual (LMS, collaboration tools, etc.)	
	Define accessibility plan	
	Cultures of Change	
	Define strategy for developing a culture of change that includes:	
	• Identification of 'transformers' – teachers to help lead change	
	• Realistic expectations set and communicated to staff	
	• Support for change hesitant staff	
	• Sufficient time to allow for gradual change	
	Develop structured Professional Learning Program:	
	• Provide sharing / collaboration facilities (online, in school, scheduling, etc.)	
	• Select technology coach(es) / integrationist(s)	

Tools and Resources

	• Design cascading plan to support teachers at different stages of change	
	• Develop a plan for ongoing, teacher-led action research	
	Schedule and begin Professional Learning (prior to distributing devices)	
	Develop financial strategy:	
	• Determine and clearly outline ownership model(s) - school owned / leased vs. parent owned / BYOD	
	• If co-contribution determine any additional sources of funding	
	• If BYOD, define minimum requirements	
	• Address equity issues around devices and connectivity (provision for financially challenged families)	
	Outline Communication/Public relations strategy that:	
	• Introduces initiative to all teachers, parents, students and community	
	• Promotes successes	
	• Helps gain ongoing broad support for program	
	Determine role and level of involvement of key stakeholders	
	Implementation	
	Readiness Assessments	
	Readiness assessment detailed audits including:	
	• Infrastructure, network, other technology	
	• Facilities	
	• Pedagogical – teacher readiness	

Tools and Resources

	Implementation Options and Project Plan	
	Detail grade/class rollout plan (cohorts, expansion, etc.)	
	Define detailed timetable of phase-in	
	Outline deployment plan (including dates when infrastructure, devices, software and apps are required)	
	Device and Software / Apps Selection	
	Analyze devices based on selection criteria, including:	
	• Functionality – does it do what is needed?	
	• Platform	
	• Range of peripherals	
	• Durability	
	• Weight	
	• Memory	
	• Speed, memory to run required applications	
	• Minimum battery life required	
	• Sufficiently future proofed to allow software / hardware upgrades	
	• Age appropriateness	
	Determine initial selection of software tools to fit pedagogical goals	
	• Explore alternative products	
	• Ensure software is robust and has future relevance	
	• Explore current licensed products	
	• Review licensing requirements for new programs/apps	

Tools and Resources

	Clarify and communicate process for additional teacher selected software implementation	
	Develop evaluation models for determining effectiveness and improvements in learning outcomes	
	Infrastructure	
	Develop a detailed infrastructure / technical upgrade plan based on determined requirements (preferred and minimum) including:	
	• Bandwidth scalability	
	• Internet / network security	
	• Cloud strategy	
	• Network topography and speed	
	• School electrical system reviewed	
	• Power supply / recharge stations	
	• Surge protectors / amperage	
	• Load balancing	
	• Device storage & security (lunches, PE, etc.)	
	• Room security	
	Upgrade technology / infrastructure based on plan	
	Budget	
	Develop detailed budget for upgrading infrastructure	
	Develop detailed budget for recurring costs including:	
	• Electricity	
	• Bandwidth	

Tools and Resources

	• Technical support staff	
	• Communications	
	• Professional Learning	
	• System maintenance and upgrading	
	Set an affordable total cost of ownership including:	
	• Service costs	
	• Insurance costs, where applicable	
	• Printing / internet costs	
	• Other	
	Critical Partnerships	
	Explore critical partnerships around infrastructure and support for:	
	• Telecommunications needs	
	• Professional learning	
	• Student learning	
	Define process to evaluate critical partnerships	
	Support Services	
	Service Level Agreement negotiated and includes:	
	• Key Performance Indicators (KPIs) defined	
	• Warranty response time defined	
	• Non-warranty response time defined	

Tools and Resources

	• Processes and procedures agreed with supplier	
	Establishment of service / support desk including:	
	• Staff assigned	
	• Location selected	
	• Spare parts / stock determined	
	• Student helpers – role defined, selection process defined, training plan	
	• Invoicing / cost center structures	
	• Processes for insurance & warranty claims	
	• Process for logging repairs, other services, re-imaging put in place	
	• Connectivity issues addressed (printers, scanners, projectors, whiteboards, etc.)	
	Build device images	
	Define process for upgrading images remotely if needed	
	Policy/Procedure Requirements	
	• Insurance – mandatory vs. optional/ school vs. home	
	• Parent training sessions - mandatory vs. optional	
	• Internet / network policy – home vs. school	
	• Email	
	• Social media	
	• Games policy	
	• Personal software policy	

Tools and Resources

	• Battery charging – student/parent responsibility, swap out batteries	
	• Devices left at home – spare devices	
	• Backup / Data storage – division of responsibility	
	• Virus protection / removal (cost of reimaging)	
	• Storage – mandatory vs. optional secure storage	
	• School based service/ support (cost, level of support)	
	• Other:	
	Outline and communicate clear guidelines on how new policy needs are addressed.	
	Prepare student handbook and Responsible Use Policy document	
	Ongoing Communications	
	Initiate communication plan:	
	• Schedule ongoing teacher meetings	
	• Schedule parent meetings	
	• Prepare newsletters, announcements	
	Prepare answers to anticipated questions from parents	
	Prepare a process for answering unanticipated questions from parents	
	• Share process with all staff	
	Schedule demonstrations and celebrations of success	

Tools and Resources

	Teacher and Student Deployment	
	Ensure network upgrades / infrastructure sufficient to support student use	
	Teacher devices ordered and deployed early including:	
	• Software installed	
	• Virtual/collaborative spaces installed/available	
	All teacher devices registered / serial numbers tracked	
	Teachers sign terms and conditions	
	Students receive policy handbook prior to laptop deployment	
	Parent information / training sessions completed	
	Student devices ordered early and on time delivery assured	
	Software installed	
	All student devices registered / serial numbers tracked	
	Student devices distributed	
	Student induction / workshops on delivery	

Articulate Your 1:1 Vision

What will success look like? It is often difficult to begin the process of defining a vision. Having your team begin with this activity will help get the process rolling. Read the following vision statements. Which statements best describe what your school/district will look like once you've successfully implemented the vision? Have each person on your team independently select three and write each one on a sticky note. Then, as a group, choose your top three.

1. **All students in our school, no matter what their background or family income, have the digital tools and connectivity they need for learning, when and where they need them, in order to benefit from the learning opportunities these tools make possible.**

2. **All learning experiences for all our students are built around collaboration, creation, communication, and inquiry-based learning.**

3. **Because of our 1:1 initiative, all graduates of our school system are future-career ready, and, as a result, there is a greater potential for the region/country to experience economic growth.**

4. **Learning for all of our young people is personalized, by which we mean _____.**

5. **In our school(s), we ensure learning regularly takes place not only in the classrooms, but beyond school walls. Learning can and does take place anywhere.**

6. **In our school(s), we continuously improve academic outcomes for all our students by making full use of the learning opportunities 1:1 enables.**

7. **All students at our school(s) have a range of ways to demonstrate their understanding because 1:1 has enabled the design of new, better assessment alternatives which has resulted in teachers being better able to support students and gauge learning.**

8. **Students at our school(s) have expanded resources for learning because 1:1 has allowed us to replace and move beyond depending on physical textbooks as the core resource.**

9. **With 1:1, we expanded and continue to expand pedagogical opportunities for our students which, in turn, allows us to have higher expectations for all students.**

10. **In our school(s), we cultivate in all our students the capacity to be self-directed learners as they learn how to be better informed and make better decisions about what they do and learn in the classroom.**

11. **By implementing 1:1, we provide new and unlimited opportunities for teachers to develop new practices to ensure all students are not only learning but learning what matters.**

Tools and Resources

Designing 1:1 Pedagogy Statements

Identifying and creating your own 1:1 pedagogy statement helps you prepare to design and implement laptop teaching and learning. Begin by reading the following examples that have been created by other laptop teachers. Next, complete the paragraph or sentence starters below.

Examples of 1:1 Pedagogy Statements

Example 1

I believe that laptops are one of the most powerful and vital 'thinking and doing' tools available for my students.

As such, I will design and deliver lessons that focus on building content skills and knowledge, collaboration and communication skills, and creativity, through a focus on standards and the use of relevant 21st century tools or resources. I will provide opportunities for my students to apply what they learn to their lives in general.

At the same time, students will use their laptops daily to construct their knowledge, to communicate and collaborate with others, to create new projects, artefacts, and understandings, and to share their learning with others. They will use their 'thinking/doing tool' to contribute to our classroom learning culture.

Example 2

I believe that all students are capable of learning, but that they have different learning styles and levels of maturity and engagement. I believe that hands-on, project-based and collaborative curriculum addresses these student differences better than other methods.

With this is mind, I will use laptops and software, interdisciplinary collaboration, field experiences, and authentic assessment and student self-reflective feedback on a regular basis to push my students toward improved skills and greater understanding.

My students will use their laptops to engage with the content and curriculum of my class as they construct and apply their knowledge individually.

Example 3

I believe technology is a powerful tool to use to model and explore mathematical concepts. While instructional, step-by-step guidance is still a necessary part of teaching math in order to build skills, there are larger mathematical questions that can be presented for students to tinker with, ideas to be tested, and relationships to be understood using their laptops.

I will provide regular opportunities for my students to construct and demonstrate the math concepts they are learning. I will push my students to apply their learning to their everyday lives. I will provide opportunities for my students to use their laptops to question, explore, and learn deeper math relationships and connections.

Tools and Resources

Students will use technology to learn about and then demonstrate to others what they have learned and that will, in turn, provide additional clarifying opportunities for each. Students will use online texts and sites to watch and replay demonstrations of math that can help them understand the patterns, skills, and techniques important to memory that help for the foundation of further mathematical study.

Example 4

I believe students in the 21st century deserve to be educated in the same environment they surround themselves in outside of school, which is a technological one. By providing students with technological devices and experiences, the students will be more adequately prepared for whatever they decide to undertake in the future.

I will provide daily opportunities for my students to explore, through the use of their laptops and interacting with each other, new information in an attempt to uncover and create new knowledge.

All students will use technology as a way to access information; further develop current ways of thinking, while also having an opportunity to unveil information and ways of doing things that they may have never considered before.

Example 5

I believe that all students can successfully use technology to facilitate the learning of mathematical strategies. These strategies can be used by the student to make informed and reasonable decisions.

To meet this goal, I will require my classes to use laptops daily to complete their course work. Technology will provide access to an abundance of new resources and processes outside of the classroom. Activities and tasks will inspire learning and model real-life experiences. Assigned coursework will demonstrate and promote digital citizenship and responsibility.

To meet this goal, students will use laptops daily to extract knowledge, and develop projects to find creative solutions to problems faced outside of the classroom. Students will demonstrate the ability to use technology to gather, evaluate, and process information with a positive attitude toward using technology that supports learning and productivity.

Example 6

I believe that technology is defining and changing our future. Therefore, student learning needs to be adapted to technology in order to prepare students for academic and real-world application.

I will provide the tools necessary for my students to access, analyze, and manipulate technology in order to enhance and individualize their education so that they are prepared for application in and outside of the classroom.

Tools and Resources

All students will generate and apply technological knowledge to discover, answer, and implement questions for academic and real-world usage.

Example 7

I believe that teachers who combine technology with music performance instruction are actively providing their students with two of the most powerful learning keys available.

I will provide daily opportunities for my band students to use their devices to self-assess their own playing and to model the digital musical resources I have provided on my website and through other channels. I will provide and immerse them in the multitude of tools available (e.g. tuners, metronomes, recordings) and be certain that they make using such tools a part of their daily regimen as musicians.

All band students will use their devices daily to practice and submit performances of sections of the music we are working on at that time. They will have their devices close at hand throughout rehearsal in order to check intonation, tempos, stylistic considerations, etc.

Designing Your Pedagogy Statement

Sentence/Short paragraph 1

Overview of your teacher beliefs about student learning (how they learn, how they can use technology/laptops/tablets as learning tools, etc.) ...

- I believe that …. _____

Sentence/Paragraph 2

Use one of the following sentence starters to articulate what this belief means for your teacher practices (what you as a teacher will do) ….

- I will…. _____

- As such, I will … _____

- With this in mind, I will …. _____

- Consequently, I will… _____

Sentence/Paragraph 3

As a result of the first two paragraphs or sentences, identify how your beliefs (pedagogy) will impact student tasks (what the students will do). You may want to choose one of the following sentence starters as you begin to write this final passage.

- At the same time, students will … _____

- Finally, my students will… _____

- My students will… _____

Tools and Resources

Two final steps

Two final steps complete this vital work of identifying your beliefs and resulting actions:

- **First, post your 1:1 pedagogy statement in an obvious location so that it can be a reminder to you on a regular basis. This may be represented in a physical copy you place in your classroom, office, or lesson plan book or it may be a virtual copy you post on your class or school website. Also, share your 1:1 pedagogy statement with your students and other teachers. Sharing your statement validates it and helps you think deeply about how it affects your work.**

- **Next, you should review and possibly revise your statement yearly. As you implement 1:1 teaching and learning year after year, you will find that the process of inquiry helps you build your understanding about how students learn, particularly when they employ this powerful learning tool, whether that be a laptop or tablet.**

AALF Policy Decisions

The following is a sample of important issues and ideas that should be discussed with technical and teaching staff, students, and parents to assist in the development of effective policies and ensure clear expectations around laptop use.

- Battery charging student / parent responsibility, swap out batteries, penalties.
- Backup / data storage – division of responsibility, home vs. school (USB drive, server, cloud, other).
- Virus protection / removal (cost of re-imaging).
- Storage – mandatory vs. optional secure storage.
- School based service / support (cost, level of support, supplier agreements).
- Transport – responsibility between home & school.
- Device model options – single vs. limited range of options.
- Service / support policies, pricing, guidelines.
- School bags – mandatory vs. optional.
- Parent training – mandatory vs. optional.
- Internet / network policy (in line with existing policy) - home vs. school.
- Data limit for downloading vs. purchasing more credit.
- Email (Webmail/email client – POP3, IMAP ?).
- Reporting lost / stolen laptops / insurance.
- Chat & Web 2.0 – allowed, banned, or restricted.
- Games/Mp3/ video files.
- Personal software policy.
- If devices left at home – spare devices, penalties.
- Printing and printing credits - school supplied vs. student purchase.

Infrastructure Questions

Area	Questions to Answer
IT network A school should have well-defined plans regarding how the Internet network should be installed. (This can be an already functioning or a new network.)	• What kind of Internet access is available? • What would be the ideal bandwidth? • What kind of topology and what active tools are considered suitable for future function? • In what ways does the IT network help serve classrooms? • How will all classrooms get internet access? • How does the school want to solve network coverage? • How will the access to wireless connections be regulated? • Does the school intend to integrate classroom workstations, teachers' computers and administration workstations into a joint network? • Does the school intend to integrate printers, photocopiers and telecommunication tools into a joint network? • What kind of network and boundary security (firewall, DMZ) solutions does the school want to apply? • Does the school want to implement content filters and regulations, and, if yes, which ones? • When and in which learning areas is Internet use allowed? • What other alternative network access is available if Internet connection scrambles? • Is there any site with which the school has to establish and maintain Intranet and telecommunication connections (e.g. IP telephony)? • To what extent does the school support secure network communication (for instance, in the event of transfer of statistical data) with education management and partner schools? • To what extent does the school support the use of free chat and communication channels (Skype, Messenger, etc.)?

Tools and Resources

Introducing ICT tools A school should have well-defined plans and ideas for the purchase, operation and maintenance of the ICT tool system.	• How are classrooms prepared for the installation of presentation equipment (e.g. interactive board, projector, notebook, etc.)? Consider network termination points, electrical plugs, free wall surfaces, projector stands, location of teachers' workstations, etc. What are the preferred types of learning areas that will be used in various classrooms (collaborative area, frontal lecturing area, experiment area, brainstorming area, etc.)? • How will the storage, use, charging and protection of portable devices be handled? • How are tools operated and maintained? • How many spare parts are to be kept in stock? • Will you have student workstations and, if yes, where will they be positioned? How is it made possible for more students to access the tools (positioning of tools, availability, and schedule)? • How are out-of-use tools scrapped, stored, or disposed?
Planning server infrastructure A school should have a concrete plan and idea for the purchase, operation and maintenance of the server infrastructure.	• Does the school prefer to maintain a server room of its own or to put an external supplier in charge (cloud computing)? • Does the school prefer to finance operations using its own resources or to involve external support? • How is the physical protection of servers and administrative workstations organized (server room, restricted area, etc.)? • Does the school have its own telephone exchange system or is an external service involved? • How are printing, copying, and scanning services organized? • Does the school prefer working with low-performance printers or working with multifunctional tools connected to the network? • Does the school have maintenance service contracts covering network devices, servers, workstations, and office tools? • Does the school want to install entry and surveillance systems? • What kind of software does the school want to apply (operation system, office program pack, server software, multimedia software, etc.)?

Tools and Resources

User skills A school should specify the skills and knowledge its teachers, students, and other users are expected to have.	• Do teachers and other users have the proper skills to use the available tools? • Have users participated in any IT security training? Do teachers understand how to protect student information when accessing new websites and apps? • Do teachers know how to create and edit digital content? • Do teachers have the skills to use the electronic administration system and LMS used by the school? If not, how will they get the skills required?
Content services The school decides on the amount of content it makes available and on how information sharing is to be coordinated.	• What kind of online content is shared by the school? In what ways (individual web server, central storage, etc.)? • If a school does not have a web site, how does it want to begin supplying data? • What expansions is the school planning in terms of sharing online content (for example, information for parents, school statistics, lesson plans, home assignments, digital learning materials, forums, online help, etc.)? • What tools are involved in central data transfer?

Tools and Resources

| **Data protection and security**

A school should have an accurate plan for the protection of significant and confidential data (privacy policy). It should also have rules and regulations as well as obligations policies. | <ul><li>Has the data inventory of the school been estimated with special focus on important and confidential data?</li><li>Have data and application administrators been appointed?</li><li>Is there anyone in charge of investigating security incidents?</li><li>What measures have been taken by the school to acquire the following protection solutions: antivirus protection of workstations and applications (servers), anti-spam, content filtration, and phishing?</li><li>How are the appearance and service of unidentified tools regulated?</li><li>What virtual groups exist and what type of protection (entitlements) do these have?</li><li>What backup strategy does the school follow? How are mailing, protected, and central contents saved and archived?</li><li>How are abuses (downloading unauthorized contents, cracking and corrupting central contents, etc.) treated?</li><li>What action plans are there to handle IT incidents?</li><li>What alternative solutions are there to ensure operation if network servers, office, and telecommunication tools fail to function or there is no electricity?</li><li>How is data and information security regulated and how are these regulations enforced?</li><li>How does the school increase security/ privacy awareness?</li></ul> |

Tools and Resources

Frequently Asked Questions About 1:1

Below is a list of 100+ questions to which you should have answers that are consistent with both the values and culture of your school and the vision that you have for your 1:1 initiative. They are the issues, ideas, and challenges that can be raised by **staff**, **students**, **parents,** the **media**, or members of the **community**.

General questions

- Will my child have to take the computer to school each day? My child already has to take a lot of bags to school anyway for sports, music, and other activities.
- Why do we need laptops, anyway? After all, the school has considerable resources and many students have access to a computer at home.
- Why not just use smartphones?
- What computer skills will be taught this year and when?
- Will total conformity in computer equipment be required?
- What software will be used?
- Will the software used on the laptops be the same as that currently being used at school? If not, when will training start?
- Will my son/daughter be in danger carrying the device to and from school?

Tools and Resources

Questions about daily usage in the classroom

- **Will you** keep us informed about the integration of laptops into the curriculum? How?
- How many classes will the laptops be used in?
- What percentage of the school day will the kids use their laptops?
- Will you evaluate learning outcomes differently?
- How will the overall program be evaluated?
- How will you train the teachers?
- Will students be working collaboratively? If so, how will they send messages and files to each other?
- Will there be a central store of information that will be available to all students? If so, will students be able to connect to the server?
- How will students connect to printers, scanners, digital cameras, and projectors?
- How will students charge their laptop batteries if the batteries run low during class?
- Will students and teachers have email accounts?
- Are you providing e-mail so that students and teachers can send files and attachments—for example, homework assignments—to students who aren't in the classroom? Or will e-mail be to restricted to in-school use only?
- What level of security will you have to regulate student browsing of the Internet?
- What level of security do you need to safeguard your network against unwanted access from the internet or other sources outside the school?
- Will there be a network over which students can communicate with teachers? If so, how will you establish a secure environment for the teachers' laptops?
- What type of network access will students have? Will students be able to change or delete files on other PCs and laptops?
- Will students be able to change hardware settings on their laptops?

Tools and Resources

Finance, interest rates, and insurance questions

- What purchase/rental plans does the school offer?
- How will the program be financed? What interest rate are we paying over the period of the loan?
- What does the insurance cover? What doesn't it cover?
- How will the payment of the deductible be handled?
- Can the school get group insurance for these laptops?
- Wouldn't my household insurance cover the laptop?
- What's the life expectancy of the laptop hardware? What about the software?

Questions that address affordability, price, and equity

- What are the pricing details and their options?
- Will the school purchase the hardware in bulk to reduce costs?
- I have twins in the grade taking this program. Is there a discount for buying two laptops?
- I would like my child to be involved in the program, but I can't afford to make the monthly payments. Is there any support for parents in my position?
- I just purchased a computer for my child at home, why should I buy another?
- I don't have access to the internet at home and can't afford it. How will my child be able to connect to the Internet outside of school?

Viruses

- What security and virus-checking procedures will be implemented?
- How often will these virus-checking programs be upgraded?
- If my child's laptop gets a virus, what should we do?

Tools and Resources

Concerns about loss of basic skills

- Aside from word processing and accessing data, what advantage is there in using computers for other areas of the curriculum, such as mathematical analysis, science, and history?
- What about handwriting? Won't my child's handwriting suffer from using a keyboard all day long?
- Won't the students be able to cheat by using the spelling checker? What effect will that have on their spelling skills?
- Don't computers isolate kids? Make them less active?

Health questions

- Is being exposed to wifi and a wireless network signal all day dangerous to my child's health? Is it an electromagnetic or radiation hazard?
- How will you ensure that the weight of my child's backpack when traveling back and forth to school won't be too great, thus creating a health risk?
- Will working on a laptop all day harm my child's eyesight?

Infrastructure questions

- How does the school plan to use the internet?
- Do students currently have access to the World Wide Web?
- Is the school library making use of the internet?
- What provisions will there be for printing?
- How are the students expected to charge their laptop batteries?

Platform issues

- Is it possible to use _____ (a different brand) computers?
- I have a (different brand/OS) printer at home. Will my child be able to use it with the laptop?
- Will my child be able to transfer files from my home computer on another platform onto these laptops?
- Why aren't you recommending Apple Macintosh (or Windows or Chromebook, etc.) computers?
- Why specify a particular brand of laptop?

Obsolescence and upgrades

- What will be done to ensure that the hardware and software get updated in a timely and cost-effective manner to keep up with developments in the technology industry?
- Will there be a change in the demands on equipment and software?
- How soon is such a change likely to take place?
- How long before the laptops and software will need to be replaced or upgraded?
- If there is a change in the required equipment, how will teachers be able to teach new computer applications if the students' equipment is out-of-date?
- Technology advances very quickly. What level of hardware and software are parents expected to buy? Will multimedia be needed? Will wireless be required?
- How long will it be before you require us to replace (or upgrade) the hardware or software?
- Who is to use the laptops?
- Will it be appropriate for students in other grade levels to use the laptops?
- When does the school intend to introduce laptops for the entire school?
- If not now, when will students in other grades begin using laptops? Aren't those other grades missing out?

Questions about security, cybersecurity, and use outside of school

- Even with the best intentions, children still lose things. How do we cope with a lost laptop?
- Student lockers are inadequate for storing laptops safely. How will the school deal with this problem? Will there be a secure place at school to store the laptop during lunch/sports/etc?
- How do we ensure the personal safety of the students carrying these valuable items in public?
- How can I protect my child from cyberbullying?
- What about the dangers of internet predators? How can my child be protected?
- Will there be filters on the laptops?
- Will these filters also be applied when the laptop is at home?
- In school, will students be allowed to use Facebook? Youtube? Other social media? Games? What about at home?
- How do I make sure my child isn't viewing inappropriate sites? Or using social media inappropriately?
- Will you help students learn about online privacy issues?
- How much data are you collecting on my child?
- How are you using this information?
- Who has access to this information at the school? Outside of the school?

Support for parents

- Do you have tips for parents on managing use of the laptops when they are at home?
- Where at home should my child be using the laptop? Is it okay if they use it while studying alone in their bedrooms?
- How do I keep my child from playing games extensively on the computer?
- How will you advise parents of very young children around technology use outside of school?
- When at home, for approximately how much time will my child be expected to be using his or her device for homework?

Support services

- Who will be servicing the laptops? How long will it take to complete repairs and have a laptop back to the student?
- How do you decide if a repair is a warranty or an insurance claim?
- Is the school going to offer any service or support facilities?
- Will we be charged an extra fee after a certain number of repairs?
- Will you use students to help with any of the technical support?
- What if I can't connect to the internet at home – will there be any support services offered by the school that I can contact for assistance?

Questions around cabling and network topology

- Will the laptops and PCs have any form of peer-to-peer connection (for example, Bluetooth) or will they connect to a server?
- Will there be a LAN (local area network) at each school site? Do you have an existing LAN to which you will be connecting the laptops?
- Which PCs and laptops will be connected to the network?
- Will there be network stations in the classroom or one stand-alone PC connected at all times?
- (If the plan includes more than one school) Will the schools be connected over a WAN (wide area network) and will there be internet filtering?
- How will you make the trade-off between increased bandwidth and increased cost of cabling?